SpringerBriefs in Molecular Science

Chemistry of Foods

Series Editor

Ricardo Pereira, Centre of Biological Engineering, University of Minho, Braga, Portugal

The series Springer Briefs in Molecular Science: Chemistry of Foods presents compact topical volumes in the area of food chemistry. The series has a clear focus on the chemistry and chemical aspects of foods, topics such as the physics or biology of foods are not part of its scope. The Briefs volumes in the series aim at presenting chemical background information or an introduction and clear-cut overview on the chemistry related to specific topics in this area. Typical topics thus include:

- Compound classes in foods—their chemistry and properties with respect to the foods (e.g. sugars, proteins, fats, minerals, …)
- Contaminants and additives in foods—their chemistry and chemical transformations
- Chemical analysis and monitoring of foods
- Chemical transformations in foods, evolution and alterations of chemicals in foods, interactions between food and its packaging materials, chemical aspects of the food production processes
- Chemistry and the food industry—from safety protocols to modern food production.

The treated subjects will particularly appeal to professionals and researchers concerned with food chemistry. Many volume topics address professionals and current problems in the food industry, but will also be interesting for readers generally concerned with the chemistry of foods. With the unique format and character of SpringerBriefs (50 to 125 pages), the volumes are compact and easily digestible. Briefs allow authors to present their ideas and readers to absorb them with minimal time investment. Briefs will be published as part of Springer's eBook collection, with millions of users worldwide. In addition, Briefs will be available for individual print and electronic purchase. Briefs are characterized by fast, global electronic dissemination, standard publishing contracts, easy-to-use manuscript preparation and formatting guidelines, and expedited production schedules.

Both solicited and unsolicited manuscripts focusing on food chemistry are considered for publication in this series. Submitted manuscripts will be reviewed and decided by the series editor, Dr. Ricardo Pereira.

Shahnshah E. Azam · Luís Loureiro ·
António A. Vicente · Renata Vardanega

Cannabinoids in Food Science

Chemistry, Applications, and Innovations

 Springer

Shahnshah E. Azam
Centre of Biological Engineering
University of Minho
Braga, Portugal

António A. Vicente
Centre of Biological Engineering
University of Minho
Braga, Portugal

LABBELS—Associate Laboratory
Braga, Portugal

Luís Loureiro
Centre of Biological Engineering
University of Minho
Braga, Portugal

LABBELS—Associate Laboratory
Braga, Portugal

Renata Vardanega
Centre of Biological Engineering
University of Minho
Braga, Portugal

LABBELS—Associate Laboratory
Braga, Portugal

ISSN 2191-5407   ISSN 2191-5415   (electronic)
SpringerBriefs in Molecular Science
ISSN 2199-689X   ISSN 2199-7209   (electronic)
Chemistry of Foods
ISBN 978-3-032-11545-4   ISBN 978-3-032-11546-1   (eBook)
https://doi.org/10.1007/978-3-032-11546-1

This Springer imprint is published by the registered company Springer Nature Switzerland AG
The registered company address is: Gewerbestrasse 11, 6330 Cham, Switzerland

If disposing of this product, please recycle the paper.

Contents

Chapter 1
Introduction

Cannabis sativa L., a member of the Cannabaceae family, is a versatile species naturally adapted to a variety of climates [1, 2]. For millennia, it has been valued not only for its undeniable medicinal properties but also as a source of raw materials for diverse applications, including textile fibers, oil, and animal feed [3]. In recent decades, interest in cannabis has grown in the food sector, largely due to its rich profile of bioactive compounds and nutrients [2]. Nevertheless, it remains the subject of considerable debate, primarily because of its psychoactive potential and variability in chemical composition among cultivars [3].

The taxonomy of cannabis plant has long been debated, driven by its remarkable genetic diversity, extensive history of cultivation, and frequent hybridization. While some taxonomic systems recognize a single, highly variable species (*Cannabis sativa* L.) with multiple subspecies or varieties [1], the most widely accepted classification distinguishes three species: *Cannabis sativa*, *Cannabis indica*, and *Cannabis ruderalis* Janisch [4]. These species differ in their morphology, growth cycles, and typical phytochemical profiles, which has important implications for both industrial and medical applications.

The terms "hemp" and "marijuana" both refer to plants of the *Cannabis sativa* L. species, but they represent different cultivars or chemotypes differentiated primarily by their Δ^9-tetrahydrocannabinol (Δ^9-THC) content. Collectively, both are encompassed under the general term "cannabis." Hemp cultivars contain low levels of Δ^9-THC, while marijuana cultivars contain higher concentrations [5]. Hemp is recognized as the oldest cultivated fiber crop, with its seeds and seed oil historically used as food sources. Marijuana, conversely, has a long tradition of both recreational and medical use [5, 6].

Cannabis terminology often causes confusion, particularly in public discourse. Marijuana is frequently perceived solely as a recreational drug, overshadowing the broader relevance of cannabis as a source of industrial materials, nutritional products, and medicinal compounds. In essence, marijuana refers to cultivars with psychoactive potential, used for both therapeutic and recreational purposes, whereas hemp refers

© The Author(s), under exclusive license to Springer Nature Switzerland AG 2026
S. E. Azam et al., *Cannabinoids in Food Science*,
Chemistry of Foods, https://doi.org/10.1007/978-3-032-11546-1_1

to non-intoxicating cultivars with diverse applications. These range from fabrics and textiles to nutraceuticals, pharmaceuticals, food and beverages, oral and topical self-care products, veterinary formulations, and a wide variety of other industrial goods [5]. This distinction is critical for policy, commerce, and public understanding.

From a phytochemical perspective, cannabis produces a wide range of bioactive compounds. Its primary metabolites—such as amino acids, fatty acids, and steroids—are essential for basic plant growth and development [3]. More notable, however, are its secondary metabolites, which include phytocannabinoids, flavonoids, terpenoids, lignans, and alkaloids. Among these, phytocannabinoids are the most extensively studied group. These terpenoid compounds are synthetized and accumulated predominantly in the granular trichomes of female inflorescences, where they serve protective roles against herbivores, pathogens, and environmental stressors. They are largely responsible for the plant's therapeutic properties, with compounds such as Δ^9-THC and cannabidiol (CBD) receiving the greatest pharmacological attention [2, 7]. Δ^9-THC is the main responsible for the psychoactive effects experienced upon inhalation or oral ingestion of cannabis, while CBD is not psychoactive.

Scientific evidence supporting the therapeutic potential of cannabis and its bioactive compounds has been expanded rapidly in recent years. Clinical and preclinical studies have demonstrated that specific cannabinoids can modulate multiple physiological pathways, offering benefits across a wide range of medical conditions. A recent systematic study classified the evidence for certain applications—such as the treatment of multiple sclerosis, epilepsy, chemotherapy-induced nausea and vomiting, pain, and appetite stimulation—as high quality, supported by robust clinical data [8]. In contrast, the evidence for potential benefits in neurodegenerative disorders, cancer diseases, psychiatric disorders, alcoholism, and dermatological conditions disorders as moderate to low quality, highlighting areas where further rigorous research is needed [8].

The health-promoting and nutritional potential of cannabis components has been promoting global interest in cannabis-derived products, with edible products emerging as one of the most promising and versatile options [5, 9–11]. Most commercially available edibles with declared cannabinoid content are formulated using hemp oil or hemp meal seeds and, in some cases, cannabis extracts [3]. Current products range from beverages (tea, coffee, beer, wine, sodas) to confectionery (lollypops, chocolates), baked goods (cookies, brownies, breads), beef jerky, honeys, etc. [10, 11].

While hemp ingredients offer notable nutritional and functional benefits, challenges remain for their use in food systems [3]. From a food science perspective, limited research has explored how ingestion compares with other modes of cannabinoid's consumption and how cannabinoids behave under industrial processing conditions [10]. Furthermore, although many studies have reported the health benefits of cannabinoids, others raise concerns about its safety, consumers risk, and potential side effects [12]. Addressing these gaps is essential for the responsible development and regulation of cannabis-based edibles in the global market. In the following chapters, this book presents a comprehensive overview of the current state of knowledge on cannabis in the field of food science, with a particular emphasis on its chemical composition, cannabinoid metabolism, and the analytical techniques used for cannabinoid detection. It also explores the diverse applications and emerging innovations involving cannabis in the food sector. Furthermore, it examines the current global regulatory landscape, highlighting the challenges and opportunities that shape this rapidly evolving field.

1.1 Historical Overview of Cannabis

Paleobotanical evidence indicates that cannabis has been present for approximately 11,700 years in Central Asia, particularly near the Altai Mountains, which many experts regard as its most likely origin. However, in 1996, the Russian botanist Nikolai Vavilov proposed that cannabis cultivation may have independently developed in several regions of the world at the same time [13]. Archaeological findings further revealed details about cannabis use in early human societies. at the Okinoshima Mesolithic site in Japan, seeds attached to broken pottery, associated with the Jomon culture (12,500–2300 BC) were discovered [14]. Additional remains from the Matsugasaki and Torihama sites, as well as a pot from the Shobuzaki Shell Midden, confirm the presence of cannabis during the period 6000–5220 BC [14]. In ancient China, cannabis was not only cultivated for industrial use, but also valued for its psychoactive properties, as evidenced by depictions of the plant on Yangshao-era pottery dating back from 6200 BC [14]. A remarkable discovery was made in Xinjiang Uyghur, where a Tocharian grave dating to 750 BC contained cannabis with a high tetrahydrocannabinol (THC) content. Analyses revealed that only the male plant parts, which contain lower THC levels, were removed, suggesting intentional cultivation for psychoactive use [15, 16].

The medical use of cannabis was also deeply rooted in early Chine culture. Around 2900 BC, Emperor Fú Xī documented "Ma"—the Chinese term for cannabis—emphasizing its broad therapeutical applications and its role in balancing yin and yang, opposing yet complementary forces central to Chinese philosophy. Even the Chinese character for "Ma" is thought to represent the drying of cannabis stems hung upside down [17, 18].

From China, the knowledge of cannabis spread westward. In India, the plant still holds profound religious and cultural significance with early references found in the Vedas, ancient Hindu scriptures. Beyond the spiritual use, cannabis was also adopted for a wide range of medicinal purposes: as an analgesic for neuralgia, headaches, and toothaches; as an anticonvulsant for epilepsy, tetanus, and rabies; as a hypnotic and sedative to treat anxiety, mania, and hysteria; and as an anesthetic and anti-inflammatory agent, especially for rheumatism. Additional benefits included antibiotic, antiparasitic, antispasmodic, diuretic, aphrodisiac, antitussive, and expectorant properties [19, 20].

This tradition of medicinal use continued as cannabis reached the Arab world, likely through traders from India. During the Islamic Golden Age, the Persian physician Avicenna described cannabis extensively in his influential work *The Canon of Medicine* (tenth century). He recommended it for headaches, joint diseases, eye inflammations, gout-related edema, wounds, and uterine pain. Avicenna's writings profoundly shaped Western medicine from the thirteenth to the nineteenth centuries [14, 21, 22].

Medieval Arabic texts provide additional evidence of cannabis as a treatment for epilepsy. The Persian physician al-Majusi prescribed cannabis leaf juice instilled into the nostrils to prevent seizures. Cannabis also entered European imagination through cultural exchange. The *One Thousand and One Nights* tales, popular between the tenth and seventeenth centuries, introduced hashish-broadly referring to resin, leaves, flower heads, and edibles, which were typically ingested rather than smoked. The word "assassin" is even believed to derive from "Hashishin," or hashish eaters, as noted by Marco Polo during his travels (1254–1324) [23–26].

Historical records also point to cannabis use in Europe and the Mediterranean before the Christian era. The Scythians, who migrated from Central Asia through Russia around 3500 years ago, played a significant role in its dissemination. Paleobotanical findings have uncovered wild cannabis pollen dating from 10,200 to 8,500 BC in Romania, Bulgaria, and Hungary [14, 19]. The Greek historian Herodotus provided detailed description of Scythian purification rituals, in which cannabis seeds were thrown onto heated stones to release dense vapors that participants inhaled [27]. Supporting these accounts, the Russian anthropologist S.I. Rudenko uncovered a Scythian burial in 1937 that contained burnt cannabis seeds in a bronze cauldron, hemp fiber shirts, and metal censers designed for inhaling cannabis smoke. His findings suggest that cannabis inhalation was a common practice in daily life, extending well beyond just religious ceremonies [17].

The introduction of cannabis to American territory remains a subject of debate, though the prevailing theory holds that it arrived in South America, particularly in Chile, during the mid-1500s. By the mid-1600s, its cultivation spread to North America, where cannabis was used in religious rituals, as medicine, and as source of fiber. In Latin America, Spanish monarchs encouraged hemp production during the colonial periods, promoting its strong and versatile fibers for textile and rope [28].

Medical cannabis entered Western medicine in the nineteenth century, largely through the work of Dr. William O'Shaughnessy, an Irish army surgeon stationed in India. In 1843, he systematically studied the plant's therapeutic effects, documenting its popular uses; testing its toxicity; and reporting successful treatments for rheumatism, hydrophobia, cholera, tetanus, infantile convulsions, and delirium tremens [13, 26, 29].

Following O'Shaughnessy's work, other physicians began to advocate for cannabis in clinical practice. By the mid-nineteenth century, the plant was included in the American Pharmacopeia, with tinctures widely prescribed for broad range of ailments, including neuralgia, tetanus, typhus, cholera, rabies, and dysentery [30, 31]. In 1889, Dr. E. A. Birch published an article in *The Lancet* recommending cannabis for the treatment of opium withdrawal symptoms. Around the same period,

Dr. J. Russell Reynolds, Queen Victoria's personal physician, endorsed its effectiveness in treating uterine bleeding, migraines, neuralgia, and epileptic spasms, while cautioning that its effectiveness was more limited for asthma, depression, and joint pain [25].

However, the medical use of cannabis began to decline in the early twentieth century. A major limitation was the absence of knowledge about its active compounds, which restricted standardizing dosing and reproducibility. At the same time, new therapeutic options—including vaccines, aspirin, chloral hydrate, and barbiturates—were increasingly adopted in medical practice. Simultaneously, the growing availability and use of opioids in the United Kingdom and United States further marginalized cannabis as therapeutic. During this period, recreational use of plant spread, particularly among Mexican immigrants in United States and various social groups in the UK, contributing to rise stigma. These factors culminated in restrictive legislation during the 1930s, effectively banning its medical use [21, 23, 32].

Despite several earlier efforts to elucidate the structure of cannabinoids, only in 1964 Prof. Rafael Mechoulam and his team isolated and identified Δ^9-THC from hashish, conclusively stablishing it as the principal psychoactive compound in cannabis accurately determining its chemical structure [33]. This landmark discovery not only resolved a long-standing scientific gap but also reignited research interest in cannabis after its decline in medical use during the early twentieth century. Since then, the integration of traditional knowledge with modern scientific methods has advanced therapeutic applications, informed safer clinical practices, and provided insights into appropriate dosing strategies aimed at maximizing benefits while minimizing risks [28].

References

1. Duczmal D, Bazan-Wozniak A, Niedzielska K, Pietrzak R (2024) Cannabinoids—multifunctional compounds, applications and challenges—mini review. Molecules 29:4923. https://doi.org/10.3390/molecules29204923
2. Fathordoobady F, Singh A, Kitts DD, Pratap Singh A (2019) Hemp (Cannabis Sativa L.) extract: anti-microbial properties, methods of extraction, and potential oral delivery. Food Rev Int 35:664–684. https://doi.org/10.1080/87559129.2019.1600539
3. Bartončíková M, Lapčíková B, Lapčík L, Valenta T (2023) Hemp-derived CBD used in food and food supplements. Molecules 28:8047. https://doi.org/10.3390/molecules28248047
4. Thomas BF, ElSohly MA (2016) The Botany of Cannabis sativa L. In: The analytical chemistry of cannabis. Elsevier, pp 1–26. https://doi.org/10.1016/B978-0-12-804646-3.00001-1
5. Salehi A, Puchalski K, Shokoohinia Y, Zolfaghari B, Asgary S (2022) Differentiating cannabis products: drugs, food, and supplements. Front Pharmacol 13. https://doi.org/10.3389/fphar.2022.906038
6. Farinon B, Molinari R, Costantini L, Merendino N (2020) The seed of industrial hemp (Cannabis sativa L.): nutritional quality and potential functionality for human health and nutrition. Nutrients 12:1935. https://doi.org/10.3390/nu12071935

7. Citti C, Braghiroli D, Vandelli MA, Cannazza G (2018) Pharmaceutical and biomedical analysis of cannabinoids: a critical review. J Pharm Biomed Anal 147:565–579. https://doi.org/10.1016/j.jpba.2017.06.003

8. Fraguas-Sánchez AI, Torres-Suárez AI (2023) Therapeutic uses of Cannabis sativa L. Current state and future perspectives. In: Current applications, approaches and potential perspectives for hemp: crop management, industrial usages, and functional purposes, pp 407–445. https://doi.org/10.1016/B978-0-323-89867-6.00010-X

9. Marangoni IP, Marangoni AG (2019) Cannabis edibles: dosing, encapsulation, and stability considerations. Curr Opin Food Sci 28:1–6. https://doi.org/10.1016/j.cofs.2019.01.005

10. Rasera GB, Ohara A, de Castro RJS (2021) Innovative and emerging applications of cannabis in food and beverage products: from an illicit drug to a potential ingredient for health promotion. Trends Food Sci Technol 115:31–41. https://doi.org/10.1016/J.TIFS.2021.06.035

11. Lindekamp N, Triesch N, Weiss M, Voß A, Rautenberg T, Rohn S, Weigel S (2025) Hemp containing breads: cannabinoid content, profiles, and thermal stability during baking. Food Chem 487:144714. https://doi.org/10.1016/J.FOODCHEM.2025.144714

12. Engeli BE, Lachenmeier DW, Diel P, Guth S, Villar Fernandez MA, Roth A, Lampen A, Cartus AT, Wätjen W, Hengstler JG, Mally A (2025) Cannabidiol in foods and food supplements: evaluation of health risks and health claims. Nutrients 17:489. https://doi.org/10.3390/nu17030489

13. Crocq MA (2020) History of cannabis and the endocannabinoid system. Dialogues Clin Neurosci 22 (2020) 223–228. https://doi.org/10.31887/DCNS.2020.22.3/mcrocq.

14. Pisanti S, Bifulco M (2019) Medical Cannabis: a plurimillennial history of an evergreen. J Cell Physiol 234:8342–8351. https://doi.org/10.1002/jcp.27725

15. McPartland JM, Hegman W, Long T (2019) Cannabis in Asia: its center of origin and early cultivation, based on a synthesis of subfossil pollen and archaeobotanical studies. Veg Hist Archaeobot 28:691–702. https://doi.org/10.1007/s00334-019-00731-8

16. Russo EB, Jiang H-E, Li X, Sutton A, Carboni A, del Bianco F, Mandolino G, Potter DJ, Zhao Y-X, Bera S, Zhang Y-B, Lü E-G, Ferguson DK, Hueber F, Zhao L-C, Liu C-J, Wang Y-F, Li C-S (2008) Phytochemical and genetic analyses of ancient cannabis from Central Asia. J Exp Bot 59:4171–4182. https://doi.org/10.1093/jxb/ern260

17. Deitch R (2003) Hemp: American history revisited: the plant with a divided history. Algora Pub

18. Jiang H-E, Li X, Zhao Y-X, Ferguson DK, Hueber F, Bera S, Wang Y-F, Zhao L-C, Liu C-J, Li C-S (2006) A new insight into Cannabis sativa (Cannabaceae) utilization from 2500-year-old Yanghai Tombs, Xinjiang, China. J Ethnopharmacol 108:414–422. https://doi.org/10.1016/j.jep.2006.05.034

19. Bauer B, Kostić V, Čekovska S, Kavrakovski Z (2016) Cannabis history and timeline. Maced Pharm Bull 62:477–478

20. Rahic O, Vranic E, Hadžiabdić J, Elezović A (2016) Hold-time stability study-a "must-do" for pharmaceutical industry sup. Maced Pharm Bull. https://www.researchgate.net/publication/349663621

21. Mahdizadeh S, Khaleghi Ghadiri M, Gorji A (2015) Avicenna's canon of medicine: a review of analgesics and anti-inflammatory substances. Avicenna J Phytomed 5:182–202

22. Attia VI (2017) Cannabis in ancient Egypt

23. Hand A, Blake A, Kerrigan P, Samuel P, Friedberg J (2016) History of medical cannabis. J Pain Manag 9:387–394

24. Booth M (2004) Cannabis: a history. Thomas Dunne Books/St. Martin's Press

25. Abel EL (1980) Marihuana. Springer US, Boston, MA. https://doi.org/10.1007/978-1-4899-2189-5

26. Stasiłowicz A, Tomala A, Podolak I, Cielecka-Piontek J (2021) Cannabis sativa L. as a natural drug meeting the criteria of a multitarget approach to treatment. Int J Mol Sci 22:778. https://doi.org/10.3390/ijms22020778

27. Zuardi AW (2006) History of cannabis as a medicine: a review. Rev Bras Psiquiatr 28:153–157. https://doi.org/10.1590/S1516-44462006000200015

28. Laaboudi FZ, Rejdali M, Amhamdi H, Salhi A, Elyoussfi A, Ahari M (2024) In the weeds: a comprehensive review of cannabis; its chemical complexity, biosynthesis, and healing abilities. Toxicol Rep 13:101685. https://doi.org/10.1016/j.toxrep.2024.101685
29. Backes M (2017) Cannabis pharmacy the practical guide to medical marijuana—revised and updated. Hachette Books
30. Groom Q, Clarke RC, Merlin MD (2013) Cannabis: evolution and ethnobotany. Plant Ecol Evol 147(2014):149–149. https://doi.org/10.5091/plecevo.2014.933
31. O'Shaughnessy WB (2025) On the preparations of the Indian Hemp, or Gunjah: cannabis Indica their effects on the animal system in health, and their utility in the treatment of tetanus and other convulsive diseases. Prov Med J Retrosp Med Sci 5 (1843) 363. https://pmc.ncbi.nlm.nih.gov/articles/PMC2490264/. Accessed 5 July 2025
32. Glen R, Boire KM (2006) Feeney, medical marijuana law. Robin Publishing
33. Gaoni Y, Mechoulam R (1964) Isolation, structure, and partial synthesis of an active constituent of hashish. J Am Chem Soc 86:1646–1647. https://doi.org/10.1021/ja01062a046

Chapter 2
Chemical Composition, Classification, and Metabolism of Cannabinoids

2.1 Chemical Composition of Cannabis

Cannabis is a polymorphic plant with an exceptionally complex chemical profile, producing a wide variety of compounds distributed throughout its flowers, leaves, stems, and seeds. To date, more than 750 distinct molecules have been identified, including over 125 cannabinoids—a unique class of terpenophenolic compounds largely specific to *Cannabis* [1, 2]. Additionally, the plant contains approximately 120 terpenes, 42 phenolic compounds, 34 flavonoids, as well as diverse classes such as alkaloids, stilbenoids, steroids, carbohydrates, polysaccharides, fatty acids and their esters, amides, amino acids, phytosterols, polyphenols, sterols, oils, waxes, and proteins [3, 4]. Early reports documented around 423 compounds in 1980, illustrating the continuous growth in knowledge about *Cannabis* chemistry [1].

This extensive phytochemical diversity contributes to the varied physiological and psychoactive effects observed across different strains and products. Moreover, several non-cannabinoid constituents participate in the so-called "entourage effect," synergistically enhancing the activity of cannabinoids [5]. The plant's chemical composition is influenced by multiple factors including geographic origin, genetic variety, developmental stage, cultivation conditions (such as nutrients, humidity, and light), harvest timing, and post-harvest storage [1].

2.2 Classification of Cannabinoids

Cannabinoids are a large group of chemical compounds that are mostly found in the cannabis plant [6]. They are considered primarily responsible for the diverse biological activities attributed to cannabis. Initially, the term referred exclusively to the typical C_{21} terpenophenolic compounds found in the plant, along with their carboxylic acids, analogues, and transformation products [3]. Over time, its definition has broadened to include a wider range of structurally diverse molecules, encompassing endogenous ligands, plant-derived compounds, and synthetic analogues, all capable of interacting with cannabinoid receptors.

Chemically, cannabinoids are oxygenated tricyclic molecules with a benzopyran structure, generally containing 21 carbon atoms, with variations in the length of the alkyl side chain attached to the aromatic ring [5, 7]. These non-polar and lipophilic compounds are classified according to their origin into three major groups: endocannabinoids, phytocannabinoids, and synthetic cannabinoids [8]. Endocannabinoids are endogenous lipids, synthesized by most members of the *Animalia* kingdom, and are mainly found in the brain, where they regulate physiological processes through receptor interaction [3, 9]. Synthetic cannabinoids, designed to mimic or enhance the effects of plant-derived compounds, can bind to cannabinoid receptors with comparable affinity [10]. Phytocannabinoids are oxygenated aromatic metabolites exclusively found in *Cannabis sativa* plant [1]. As these groups differ in their origin, they contain different chemical structures and properties [11].

2.2.1 *Endocannabinoids*

The term endocannabinoids refers to lipid-based signaling molecules produced within the body, primarily from phospholipids found in cell membranes [12]. Endocannabinoids function in a retrograde signaling manner, typically synthesized "on-demand" from membrane precursors in response to physiological triggers such as calcium influx or neuronal activity. This distinguishes them from classical neurotransmitters, which are stored in vesicles. The endogenous nature of these compounds underpins their classification as body-derived cannabinoids, and their actions are highly localized and time-sensitive due to their rapid synthesis and degradation mechanisms [13, 14].

Unlike most phytocannabinoids, they do not contain a terpenophenolic chemical structure, and the major endocannabinoids like anandamide (AEA) and 2-arachidonoylglycerol (2-AG) are made up of long-chain polyunsaturated fatty acids [15]. AEA is produced by the combination of arachidonic acid with ethanolamine, whereas 2-AG is an ester of glycerol and arachidonic acid (Fig. 2.1) [16]. These cannabinoids are produced within the body according to need from membrane phospholipid precursors [16]. They act as short-lived signaling molecules and play a crucial role in modulating physiological processes, including pain sensation, mood regulation, appetite, and immune response [17].

Endocannabinoids can be classified along several interconnected dimensions, with chemical structure being a primary axis. Both AEA and 2-AG are derivatives of arachidonic acid, a polyunsaturated omega-6 fatty acid that serves as a precursor for various bioactive lipids. AEA is classified as a fatty acid amide, specifically N-arachidonoylethanolamine and is produced from N-arachidonoyl phosphatidylethanolamine (NAPE) through several enzymatic pathways, including

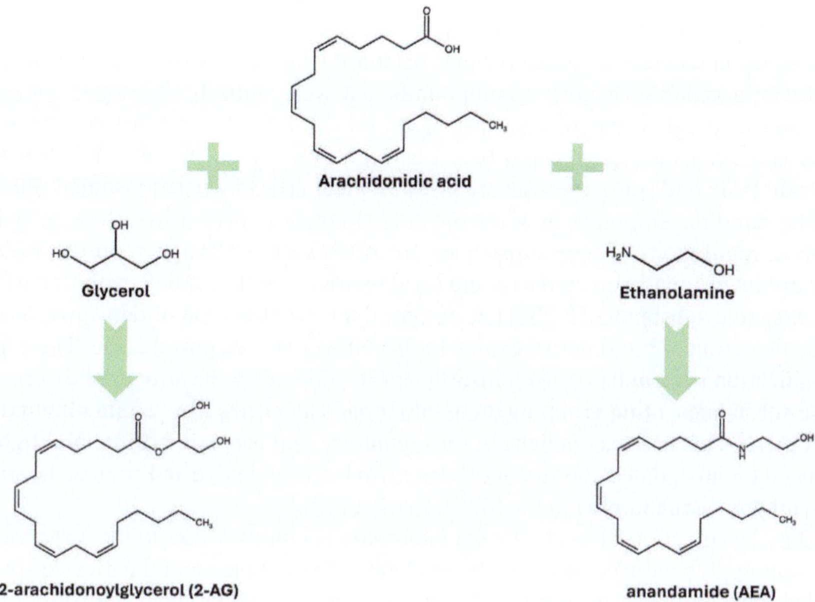

Fig. 2.1 Chemical synthesis of endocannabinoids anandamide (AEA) and 2-Arachidonoylglycerol (2-AG)

NAPE-specific phospholipase D [18, 19]. 2-AG, on the other hand, is a monoacylglycerol mainly produced from diacylglycerol (DAG) via diacylglycerol lipase (DAGL). Its structure features an ester bond between glycerol and arachidonic acid, making it chemically distinct from AEA [19].

2.2.2 Synthetic Cannabinoids

Synthetic cannabinoids are chemically designed and synthesized to imitate the effects of natural cannabinoids found in the cannabis plant and comprise a wide variety of compounds, including classical cannabinoid analogues like HU-210, non-classical cannabinoids such as WIN 55,212-2, and more recent compounds featuring indole, indazole, or other heterocyclic core structures [20]. Initially developed for research and medical purposes, many of these compounds quickly entered the illegal drug market.

These synthetic cannabinoids often contain modified hydrophobic groups, flexible linker elements, and varied alkyl side chains to precisely modulate receptor affinity, potency, and selectivity for cannabinoid receptors, including CB1 and CB2 [21].

Several synthetic cannabinoids, such as nabilone and dronabinol, have been developed and approved for therapeutic, primarily to manage nausea associated with chemotherapy or AIDS, and are being investigated for potential use in treating various types of pain, including neuropathic, spasticity-related, and nociplastic pain syndromes such as fibromyalgia, osteoarthritis, and postoperative pain [22]. However, a group of synthetic compounds that were initially developed for scientific and medical research have also been misused for illicit purposes. Many of these substances are engineered to act as high-potency agents, which can lead to unpredictable and sometimes severe physiological effects when consumed [20, 23].

The rapid development of synthetic cannabinoids is driven by efforts to bypass existing regulations. These substances are often called "designer drugs" because they are intentionally created to avoid legal restrictions. The quick evolution of new psychoactive substances (NPSs) has outpaced the development of new laws, leading to the detection of hundreds of cannabinoids NPSs over the past decade. The current classification system has proven insufficient in addressing the structural diversity of these substances, often grouping them into broad categories like "aminoalkylindole" or "other." This creates challenges for regulatory and forensic efforts, highlighting the need for an updated classification system to better organize and control the spread of synthetic cannabinoids in the illegal market [10].

The classification of synthetic cannabinoids is complex due to the large variety of compounds produced, their evolving chemical structures, and the diverse sources of these substances. Currently, a classification based on chemical scaffold offers a more precise method for categorizing synthetic cannabinoids. The chemical scaffold influences the molecule's receptor binding properties, potency, and metabolism, as well as the pharmacological and toxicological profiles of these substances [24]. Several major chemical classes have been identified among synthetic cannabinoids, as follows:

Classical Cannabinoids: Classical cannabinoids are compounds derived from dibenzopyran, including THC, its isomers, and structurally related synthetic analogs such as HU-210 [10]. HU-210 is a powerful synthetic cannabinoid that binds more strongly to CB1 receptors than THC, activating G proteins and affecting internal cell processes such as calcium signaling and glucose metabolism. Its lipophilic property helps it cross the blood–brain barrier efficiently [25].

Naphthoylindoles: This was one of the earliest and most commonly used scaffolds in synthetic cannabinoids, represented by compounds such as JWH-018 and JWH-073. These substances feature an indole core attached to a naphthoyl group and act as full agonists at CB1 receptors, with greater potency than THC. Products containing these compounds were widespread in early herbal use blends [10, 26].

Phenylacetylindoles: These compounds were developed after naphthoylindoles to better understand how their structures influence activity. They resemble naphthoylindoles but have a phenylacetyl group instead of a naphthoyl group. JWH-250 is a common example. Most of these compounds mainly target CB2 receptors, but some, like JWH-250 and JWH-203, strongly bind to both CB1 and CB2. Others, such as JWH-251, prefer CB1. All act as full CB1 agonists, although their potency and effects vary, aiding researchers in understanding their biological activity [10, 27].

Indazole and Indole Carboxamides: This class includes synthetic cannabinoids with indazole or indole cores and carboxamide groups. Indole-3-carboxamides and indazole-3-carboxamides are groups of synthetic cannabinoids that have been widely distributed as new psychoactive substances (NPS). These compounds often exhibit high affinity for cannabinoid receptors and are associated with narcogenic potential. The indole-3-carboxamide group includes compounds like CBM-018, ACBM-2201, MEPIRAPIM, and MMB-2201, all showing high affinity for cannabinoid receptors. Some of these compounds are water soluble, which adds medical value. They have been distributed as NPS since 2012 [10]. The indazole-3-carboxamides group emerged in 2012 and includes synthetic compounds such as ACBM(N)-018, ACBM(N)-2201, and MDMB(N)BZ-F. They are chemically modified to evade legal restrictions and are widely used as designer drugs [10].

Cyclohexyl Phenols: Cyclohexyl phenols, such as CP 47,497 and its homologs, are synthetic cannabinoids classified as non-classical cannabinoids. These compounds differ in structure from traditional cannabinoids but generally exhibit similar activation at the CB1 receptor [10]. This chemical scaffold classification is crucial for understanding the pharmacology and toxicology of synthetic cannabinoids, as small changes in the scaffold can greatly affect potency, receptor binding affinity, and duration of action, resulting in unpredictable effects in users [28, 29].

2.2.3 Phytocannabinoids

In the phytocannabinoids group, all the compounds have a unique terpenophenolic chemical structure, which provides the group with some distinct characteristics [30]. The terpenophenolic chemical structure is mainly composed of two parts, the monoterpene and the phenolic aromatic ring [31]. Monoterpene is a small molecule consisting of 10 carbon atoms, while a phenolic ring is a ring-shaped molecule with some specified characteristics [3]. In the terpenophenolic structure, these two compounds are joined together and form a cluster of 21 carbon atoms, usually found in major cannabinoids such as CBD and Δ^9-THC [3]. The phenolic ring usually carries a hydroxyl group (–OH) or any other group at the 1 and 3 positions of the ring. Any change in the positions or groups can develop a new type of cannabinoids, and it leads to the development of a diverse range of cannabinoid compounds [32]. These changes also alter the interaction of cannabinoids with the endocannabinoid system; hence, they also affect the physiological processes of the body, like pain, mood, appetite, and immune response [32].

In the plant, cannabinoids mainly exist as their acidic forms, such as Δ^9-tetrahydrocannabinolic acid (Δ^9-THCA) and CBDA. These acidic forms serve as the biosynthetic precursors to the neutral cannabinoids Δ^9-THC and CBD, respectively. The conversion from acidic to neutral forms usually occurs through a process called decarboxylation, which can be triggered by heat during drying, curing, or consumption [33]. The acidic and neutral forms not only differ chemically but also in their biological activity. For example, Δ^9-THCA is non-psychoactive but converts to psychoactive Δ^9-THC during decarboxylation. Likewise, the acidic forms may have different therapeutic effects and stability profiles compared to their neutral versions [34, 35]. More than 100 cannabinoids have been identified from *Cannabis*.

While impressive, this catalogue does not necessarily reflect the metabolic capacity of this genus but may be a consequence of oxidative instability of plant-derived cannabinoids, causing a range of degradation products to occur after synthesis [36]. Phytocannabinoids are generally classified into major and minor cannabinoids based on how much they are found in the plant. The most well-known major cannabinoids include Δ^9-THC and CBD, while minor cannabinoids such as cannabigerol (CBG), cannabinol (CBN), and cannabichromene (CBC) are present in smaller quantities but are gaining more interest due to their unique pharmacological effects [37, 38]. Additionally, phytocannabinoids have been subdivided into 11 categories (Fig. 2.2). The main structural characteristics of the subgroups are presented below.

Δ^9-tetrahydrocannabinol (Δ^9-THC): The chemical structure of Δ^9-THC, the principal psychoactive constituent of cannabis, was first described in 1942 and later accurately resolved in 1964 [2]. This lipophilic and volatile cannabinoid is largely responsible for the euphoric effects associated with the plant [3]. Its abundance, however, differs substantially among plant tissues [39], with flowers containing the highest proportion (10–12%), followed by leaves (1–2%), while stems (0.1–0.3%) and roots (<0.03%) exhibit only trace levels [1]. In total, 23 compounds are classified within the Δ^9-THC type. In 1964, Goani and Mechoulam successfully obtained pure Δ^9-THC after isolating it from a hexane extract of hashish. The purification process involved sequential column chromatography, first on Florisil and subsequently on alumina [2]. The isolation of $(-)$-Δ^9-trans-tetrahydrocannabinolic acid A (Δ^9-THCAA) was first achieved in 1967 through chromatography on cellulose powder, using a solvent system composed of hexane and dimethylformamide [40]. Subsequent methodological refinements optimized the extraction of both Δ^9-THC and Δ^9-THCAA from *C. sativa* material. In addition, the purification of Δ^9-THC was significantly enhanced by the application of fractional distillation [3]. $(-)$-Δ^9-trans-tetrahydrocannabinolic acid B (Δ^9-THCAB) was obtained from hashish through chromatographic separation on a silicic acid column [41]. Later, GC–MS analysis of ethyl acetate extracts derived from seized cannabis materials, including resins, tinctures, and leaves, enabled the identification and characterization of $(-)$-Δ^9-trans-tetrahydrocannabinol-C4 (Δ^9-THC-C4) and its corresponding carboxylic acid derivative, Δ^9-THCAA-C4 [42]. The cannabinoid $(-)$-Δ^9-trans-tetrahydrocannabivarin (Δ^9-THCV) was first reported in 1971 from a Pakistani cannabis tincture [2]. In 1973, the presence of $(-)$-Δ^9-trans-tetrahydrocannabivarinic acid (Δ^9-THCVAA) was documented following its isolation from fresh leaves of *C. sativa* collected in South Africa [3]. During the same year, $(-)$-Δ^9-trans-tetrahydrocannabiorcol (Δ^9-THCO or Δ^9-THC1) was identified in Brazilian marijuana through the analysis of a light petroleum ether extract [43]. Further investigations using GC–MS on various cannabis samples led to the detection of $(-)$-Δ^9-trans-tetrahydrocannabiorcolic acid (Δ^9-THCOAA) [42].

The isolation of $(-)$-Δ^9-trans-tetrahydrocannabinal (Δ^9-THC aldehyde) was reported in 2015 from a high-potency strain of *C. sativa* [44]. Several years earlier, in 2008, eight cannabinoid ester derivatives were purified from the hexane extract of the same cannabis variety using multiple chromatographic procedures. These esters were characterized as α-fenchyl $(-)$-Δ^9-trans-tetrahydrocannabinolate, epi-bornyl $(-)$-Δ^9-trans-tetrahydrocannabinolate,

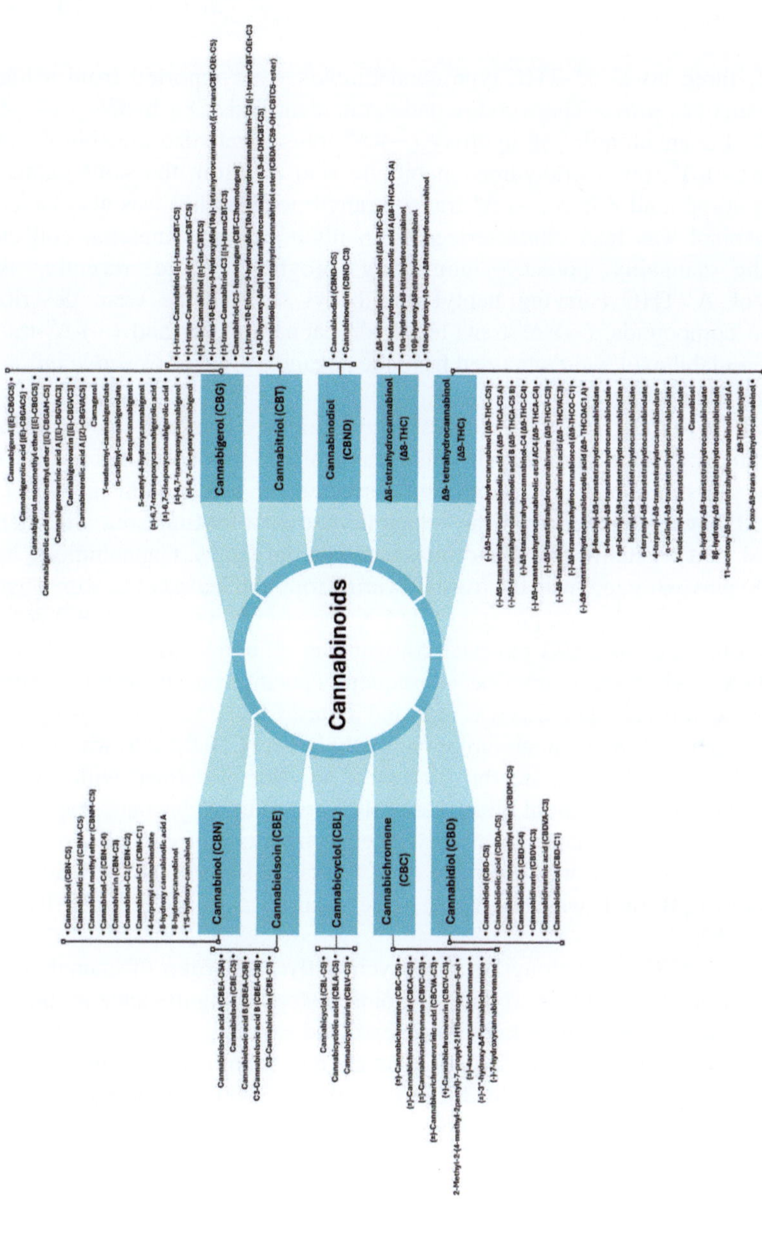

Fig. 2.2 Subcategories of cannabinoids and the compounds identified to date, with their commonly used abbreviations

β-terpenyl   (−)-Δ^9-trans-tetrahydrocannabinolate,   β-fenchyl   (−)-Δ^9-trans-tetrahydrocannabinolate,   bornyl   (−)-Δ^9-trans-tetrahydrocannabinolate,   β-cadinyl   (−)-Δ^9-trans-tetrahydrocannabinolate,   4-terpenyl   (−)-Δ^9-trans-tetrahydrocannabinolate,   and   δ-eudesmyl   (−)-Δ^9-trans-tetrahydrocannabinolate [45].

In 2015, three novel Δ^9-THC-type cannabinoids were reported from a high-potency strain of *C. sativa*. These compounds were identified as 8α-hydroxy-(−)-Δ^9-trans-tetrahydrocannabinol, 8β-hydroxy-(−)-Δ^9-trans-tetrahydrocannabinol, and 11-acetoxy-(−)-Δ^9-trans-tetrahydrocannabinolic acid A. From the same cannabis variety, the compound 8-oxo-(−)-Δ^9-trans-tetrahydrocannabinol was also isolated [46]. Cannabisol was later characterized from illicit cannabis material collected through the cannabis potency monitoring program. More recently, two homologs of Δ^9-THC carrying heptyl and hexyl side chains were described [44]. These compounds, (−)-Δ^9-trans-tetrahydrocannabiphorol and (−)-Δ^9-trans-tetrahydrocannabihexol, were obtained from the hexane extract of *C. sativa* inflorescences cultivated in Italy [47].

Cannabidiol (CBD): The CBD sub-category currently comprises seven identified compounds. CBD is a non-psychoactive cannabinoid first described in 1940 by Roger, with its structure later revised and corrected by Mechoulam in 1963 [1]. Initially, CBD was obtained from an ethanolic extract of Minnesota wild hemp and purified as a crystalline bis-3,5-dinitrobenzoate derivative. Cannabidiolic acid (CBDA-C5) was extracted from the fresh flowering tops and leaves of *C. sativa* using benzene and identified by comparison of its UV spectrum with that of CBD derivatives [3]. From Thai *Cannabis* material, Shoyama et al. extracted cannabidivarinic acid (CBDVA) via benzene extraction, subsequently purified on a polyamide column [43]. Cannabidivarin (CBDV-C3) was reported in hashish and separated using silica gel chromatography [48]. Cannabidiol monomethylether (CBDM-C5) was obtained from hemp leaves following decarboxylation of an ethanol extract, with purification achieved through sequential Florisil and silica gel column chromatography. The compound CBD-C4 originated from ethyl acetate fractions of cannabis resin and leaves, where its isolation was preceded by a derivatization step [43]. In addition, cannabidiorcol (CBD-C1) was recognized in the hexane fraction of Lebanese hashish by GC–MS [39].

Cannabinol (CBN): It belongs to the psychoactive subgroup of cannabinoids, with 11 compounds identified to date. It holds historical significance as the first naturally occurring cannabinoid to be discovered and isolated, reported by Robert S. Cahn in 1896 [49]. By 1980, the isolation details and chemical structures of seven CBN derivatives were documented. More recently, in 2009, 8-hydroxycannabinol and 8-hydroxycannabinolic acid A were isolated from a high-potency variety of *Cannabis sativa* and structurally characterized using NMR and high-resolution mass spectrometry (HR-MS) [50]. From the same plant variety, 10S-hydroxycannabinol and 4-terpenyl cannabinolate were obtained, with their structures confirmed by gas chromatography–mass spectrometry (GC–MS) [44].

Cannabielsoin (CBE): This subgroup, currently comprising five known compounds, was firstly documented in 1973. Initial detection of CBE-C5 occurred in an ethanolic extract of Lebanese hashish, which had been processed using counter-current distribution and analyzed via GC–MS [51]. One year later, its stereochemistry was determined as 5aS, 6S, 9R, and 9aR [52]. Subsequent investigations led to the isolation of cannabielsoin acid A (CBEA-C5A) and cannabielsoin acid B (CBEA-C5B) from hashish, with their molecular structures elucidated through NMR analysis and targeted chemical modifications [53]. Additional derivatives, including CBE-C3 and cannabielsoic acid B-C3 (CBEA-C3B), were later identified in cannabis plants [53].

Cannabicyclol (CBL): This subgroup comprises three identified compounds. The initial isolation of CBL from hashish was achieved by Korte and Sieper using thin-layer chromatography in 1964, although its chemical structure was accurately later resolved by Mechoulam and Gaoni through spectral analysis [54]. Cannabicyclolic acid (CBLA) was later extracted from *Cannabis* using benzene, obtained in methylated form, and determined to be an artifact arising from the natural photodegradation of cannabichromenic acid (CBCA) during storage [55]. Another related compound, cannabicyclovarin (CBLV) was identified in the ether extract of Congolese marijuana through GC–MS and GLC analysis. The formation of CBLA via irradiation of CBCA further established that CBLA is not a naturally occurring cannabinoid [56].

Cannabichromene (CBC): This comprises nine identified cannabinoids and is characterized as non-psychoactive, typically occurring in minor concentrations within the plant. Its discovery dates back to 1966, when it was first reported by Mechoulam et al. [57]. Cannabivarichromene (CBCV) was first detected by GC–MS analysis, while the related compounds CBCV and its acidic form, cannabichromevarinic acid (CBCVA), were obtained from a benzene extract of a Thai cannabis variety [58]. Additionally, three derivatives of cannabichromene, the (±)-4-acetoxycannabichromene, (±)-3″-hydroxy-Δ4″-cannabichromene, and (−)-7-hydroxycannabichromane, were purified from high-potency *C. sativa* through a combination of silica gel vacuum liquid chromatography (VLC), normal-phase silica High-Performance Liquid Chromatography (HPLC), and reverse-phase silica HPLC. In 1984, a CBC-C3 derivative was identified, and its molecular structure, the 2-methyl-2-(4-methyl-2-pentyl)-7-propyl-2H-1-benzopyran-5-ol, was elucidated through MS analysis [50].

Cannabigerol (CBG): A total of 16 cannabinoids are categorized within the CBG subgroup. Cannabigerol ((E)-CBGC5) was first obtained in 1964 from cannabis resin through Florisil column chromatography, with its chemical structure subsequently verified by total synthesis [59]. In 1975, both cannabigerolic acid ((E)-CBGAC5) and its monomethyl ether derivative ((E)-CBGC5) were isolated, confirming (E)-CBGAC5 as the primary biosynthetic precursor of Δ^9-THCAA [58]. From high-potency *C. sativa*, two (E)-CBGAC5 ester derivatives, the γ-eudesmyl cannabigerolate and α-cadinyl cannabigerolate, were separated using chiral HPLC [45]. The monomethyl ether of (E)-CBGC5 was later recovered from benzene extracts of hemp, while cannabigerovarin ((E)-CBGVC3) and its corresponding acid form ((E)-CBGVAC3) were identified in extracts of Thai cannabis [60]. Cannabinerolic acid ((Z)-CBGAC5) was obtained from acetone extracts of leaves from a Mexican *C. sativa* strain through silica gel column chromatography in 1995 [61]. From the inflorescences of high-potency *C. sativa*, 5-acetyl-4-hydroxy-cannabigerol was later separated using HPLC [50]. In 2008, four epoxy derivatives of CBG, ($\pm$)-6,7-trans-epoxycannabigerolic acid, ($\pm$)-6,7-cis-epoxycannabigerol, ($\pm$)-6,7-cis-epoxycannabigerolic acid, and ($\pm$)-6,7-trans-epoxycannabigerol were purified from high-potency *C. sativa* cultivated in Mississippi using a combination of chromatographic methods [46]. A polar dihydroxy-CBG analogue, known as camagerol, was isolated from the aerial parts of the C. sativa. Hydrolysis of the aerial waxes from this strain yielded a fraction containing sesquicannabigerol, a farnesyl-substituted CBG derivative [62].

Cannabitriol (CBT): This subgroup, comprising nine identified analogues, was first described in 1966 from Japanese cannabis, although its chemical structure was only clarified a decade later, in 1976 [63]. The CBT-type group includes (+)-trans-CBT-C5, (−)-trans-CBT-C5, ($\pm$)-trans-CBT-C3, ($\pm$)-cis-CBT-C5, a CBT-C3 homologue, (−)-trans-CBT-OEt-C3, (−)-trans-CBT-OEt-C5, 8,9-di-OH-CBT-C5, and the CBDA-C59-OH-CBT-C5-ester, all of which have been isolated from *Cannabis* species [3]. Stereochemical configuration was subsequently confirmed through X-ray crystallography. Among these, (+)-trans-CBT-C5 and (−)-trans-CBT-OEt-C5 were obtained by Elsohly et al. [64] from ethanolic extracts of cannabis. The presence of the two ethoxy derivatives is believed to be artifacts, arising from ethanol use during extraction [65].

Cannabinodiol (CBND): CBND represents a psychoactive cannabinoid subclass comprising two known compounds. The first, cannabinodivirin (CBND-C3), was identified in 1972 in hashish through GC–MS analysis [3]. Five years later, in 1977, cannabinodiol (CBND-C5) was obtained from Lebanese hashish via silica gel column chromatography. Structural elucidation of CBND-C5 was achieved using 1H-NMR spectroscopy and further validated through the phytochemical conversion of cannabinol into CBND [66].

Δ^8-tetrahydrocannabinol (Δ^8-THC): In 1966, the Δ^8-THC was first isolated from the flowers and leaves of *Cannabis* cultivated in Maryland. Purification from the petroleum ether extract was achieved using silicic acid column chromatography [67]. Nearly a decade later, the corresponding carboxylic acid, Δ^8-trans-tetrahydrocannabinolic acid A (Δ^8-THCA), was isolated as its methyl ester from a *Cannabis* specimen originating from Czechoslovakia [68]. More recently, in 2015, three hydroxylated derivatives of the Δ^8-THC type were reported from high-potency *C. sativa*. Their structures, the 10α-hydroxy-Δ^8-tetrahydrocannabinol, 10β-hydroxy-Δ^8-tetrahydrocannabinol, and 10a-α-hydroxy-10-oxo-Δ^8-tetrahydrocannabinol, were elucidated through detailed 1D and 2D NMR spectroscopic analyses [44].

Miscellaneous: Cannabinoids represent a heterogeneous group of compounds that do not fall within the major established classes. These molecules are generally detected in low abundance and often display uncommon structural features, including atypical side-chain lengths, unusual oxidation states, or acetylated and hydroxylated derivatives. In several cases, they have been reported as rare natural constituents of Cannabis sativa, while in others they may arise as artifacts of extraction or isolation procedures [3].

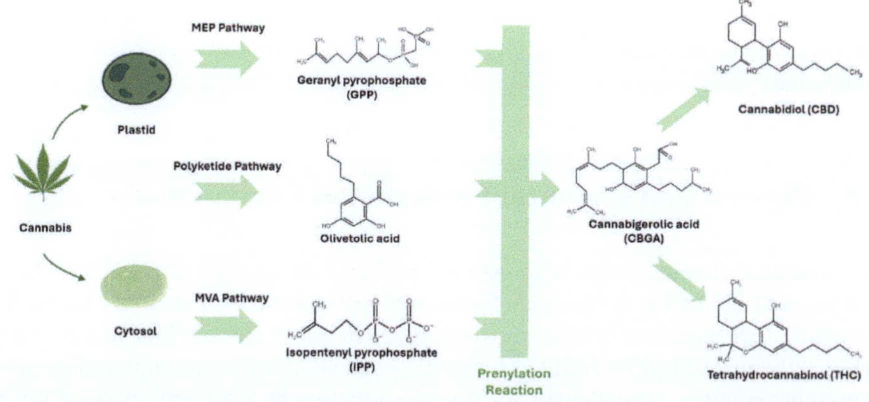

Fig. 2.3 Representation of the biosynthesis of the major phytocannabinoids cannabidiol (CBD) and Δ^9-tetrahydrocannabinol (Δ^9-THC)

As illustrated in Fig. 2.3, in *Cannabis sativa* L., the terpene portion of cannabinoids is predominantly synthesized from geranyl pyrophosphate (GPP), a key intermediate in terpenoid biosynthesis [69]. GPP is mainly produced within plastids via the methylerythritol phosphate (MEP) pathway, which starts from 1-deoxy-D-xylulose 5-phosphate and proceeds through several intermediates, including 2-C-methyl-D-erythritol 2,4-cyclodiphosphate (MEcDP) [70]. In parallel, the mevalonate (MVA) pathway, operating in the cytosol, contributes to the synthesis of other terpenoid classes such as sesquiterpenes and larger isoprenoids. This pathway generates the universal five-carbon building blocks, isopentenyl pyrophosphate (IPP) and dimethylallyl pyrophosphate (DMAPP), which can also feed into GPP biosynthesis [71]. The GPP molecules undergo prenylation with the polyketide-derived olivetolic acid, yielding cannabigerolic acid (CBGA)—the universal biosynthetic precursor of major cannabinoids such as Δ^9-THCA, CBDA, and CBCA [71].

The chemical diversity of phytocannabinoids arises from differences in three main components: the isoprenyl group, the resorcinol core, and the side chain. These variations are largely independent of each other and do not share a common biosynthetic origin. In addition, diversity is introduced through biochemical processes such as oxidative cyclization, enzymatic modifications, and other alterations that occur during their biosynthesis [36]. These structural variations lead to the formation of the various classes of phytocannabinoids (Fig. 2.2) [36]. For instance, Δ^9-THC features a partly saturated cyclohexene ring, interacts with CB1 receptors in the central nervous system, and is responsible for psychoactive effects [72]. In contrast, CBD has an open-ring structure; engages with multiple receptors; and exhibits anxiolytic, anti-inflammatory, and neuroprotective properties, without psychoactive properties [73]. CBG serves mainly as a precursor to CBD and Δ^9-THC, playing an important role in cannabinoids formation [74] while CBC is associated with anti-inflammatory and antimicrobial properties [75]. Oxidative degradation of Δ^9-THC produces CBN, which contributes to sedative effects [76]. Even minor alterations in the chemical structure of cannabinoids can significantly influence their biological activity, explaining the wide spectrum of therapeutic potentials observed across these compounds [77].

2.3 Physical and Chemical Properties of Cannabinoids

The structural characteristics of cannabinoids highly affects their physical and chemical properties. Owing to their rich hydrocarbon structure, cannabinoids are highly lipophilic and therefore exhibit poor solubility in water [78]. While this property enables them to dissolve readily in organic solvents such as ethanol and hexane, it presents significant challenges for drug formulation and delivery, particularly in oral and aqueous-based products [78]. The presence of polar functional groups, such as hydroxyl and carboxyl moieties, imparts slight polarity and modestly improves solubility, particularly in acidic cannabinoids like Δ^9-THCA and CBDA [79]. As the pH increases, the carboxylic groups of these acidic cannabinoids undergo deprotonation, significantly enhancing their aqueous solubility. At higher pH, they predominantly exist in their single anion form, which is associated with smaller droplet size and greater permeation rates in microemulsion systems. The higher diffusivity of these ionized microspecies significantly enhances flux to the receptor compartment, highlighting the critical role of pH in the solubility and permeability of acidic cannabinoids [80].

Another important physicochemical parameter is the volatility. Boiling point is generally reported as the temperature at which a compound boils under atmospheric pressure (the normal boiling point) [81]. However, considerable misinformation persists regarding the boiling points of cannabinoids and certain terpenes. For instance, many sources state that Δ^9-THC boils at 155–157 °C and CBD between 160 and 180 °C [82].

Solid-state properties also play a critical role in cannabinoids formulation and stability. CBD demonstrates crystalline characteristics, crystallizing in the monoclinic space group P21, where hydrogen bonding and supramolecular interactions stabilize the molecular framework. Within the crystal unit, two distinct conformers exist, differing mainly in the orientation of their unsaturated alkyl side chains. Both strong and weak hydrogen bonds, along with C–H⋯O and C–H⋯π interactions, contribute to maintaining the crystal integrity. The crystalline form improves stability, whereas the amorphous form, despite offering higher bioavailability, is less stable and prone to pressure-driven crystallization due to the rearrangement of weaker hydrogen bonds [83]. By contrast, Δ^9-THC generally exists as a resinous or amorphous substance with lower structural order [84]. Furthermore, cannabinoids can exhibit polymorphism, existing in multiple crystalline forms with distinct melting points, solubility, and dissolution characteristics [84]. Such polymorphic variation significantly affects bioavailability, formulation performance, and storage stability of cannabinoids [84].

Finally, partition coefficients (log P value) provide information about the lipophilic-hydrophilic behavior of cannabinoids. High log P values, such as 6.3 for CBD, confirm their strongly lipophilic character [85]. This lipophilicity facilitates dissolution in fats and oils and promotes absorption across biological membranes. At the same time, it contributes to their pharmacokinetic behavior, favoring accumulation and prolonged retention in adipose tissues [86].

2.4 Interactions Between the Functional Groups of Cannabinoids and Receptors

The type and position of different functional groups play an important role in determining the chemical, pharmacological, and metabolic characteristics of cannabinoids [87]. The most important functional group is the hydroxyl group of the aromatic ring. It commonly interacts with the receptors through hydrogen bonding and plays an important role in initiating the metabolic activity of the cannabinoids [88]. For example, in Δ^9-THC, there is a single hydroxyl group in the aromatic ring, which binds to the receptor CB1 and is responsible for its psychoactive effects [89]. While CBD has two hydroxyl groups in the aromatic ring, it can interact with different receptors, and therefore CBD has a wide range of therapeutic potential [90]. The hydroxyl groups can be chemically modified through processes like methylation or acetylation, which can alter the receptor affinity and bioavailability of the cannabinoids [89].

The alkyl side-chain group is also very important. Usually, in most cannabinoids, it is a pentyl (five-carbon) chain. It affects lipophilic behavior and influences the receptor's binding interactions [91]. The difference in chain length of the alkyl group also affects the strength of cannabinoids interaction of cannabinoids with receptors [92]. For example, cannabinoids with shorter chains, such as tetrahydrocannabivarin (THCV), reduce their activity at CB1 receptors, while longer chains, such as in tetrahydrocannabiphorol (THCP), enhance their binding affinity and strength [92]. In addition, branching and substitution of the alkyl side chain also change the receptor selectivity and functional activity of the cannabinoids [72].

The presence and position of the double bond, particularly in the cyclohexene ring, are also very important in determining the characteristics of the cannabinoids [93]. For example, the position of the double bond at the Δ-9 position in THC is responsible for its psychoactive characteristics, and the shifting of this bond's location produces new isomers like Δ^8-THC, which exhibit slightly altered pharmacological effects [93]. If this double bond is finished, such as in dihydro-THC, it significantly diminishes its psychoactivity [91].

The presence of a carbonyl functional group is not common in usual cannabinoids; they are mostly found in oxidized metabolites such as CBN, which is often produced from the oxidation of THC [94]. The presence of a carbonyl functional group generally affects molecular stability and reduces the psychoactive potential [94]. The synthetic cannabinoids, during their synthesis, include additional polar groups such as carbonyls, amides, and halogens to enhance receptor binding, improve selectivity, and optimize pharmacokinetic properties [95]. Functional groups like esters and ethers are also added in synthetic analogs to increase solubility, metabolic stability, or resistance to enzymatic degradation [95]. The changes in the functional group and their positions developed a variety of cannabinoids. THC and its analogs have different isomerization, side-chain length, and saturation levels. CBD and its analogs also vary in side-chain length and oxidation [96].

2.5 Cannabinoids Absorption, Distribution, and Metabolism

2.5.1 Absorption

Before bioactive compounds can be absorbed by the human body, they must first be released from the food matrix during digestion and become available for uptake by the intestinal epithelial cells [97]. This fraction of a compound that is liberated during digestion and is accessible for absorption is referred to as bioaccessibility. For hydrophobic compounds such as cannabinoids, absorption requires incorporation into mixed micelles, which act as transport vehicles delivering them to the intestinal epithelium. These micelles are formed through the action of endogenous surfactants (bile salts and phospholipids) together with exogenous surfactants derived from the food matrix, including free fatty acids and monoglycerides. When the dietary oil phase is digestible (e.g., vegetable oils), lipases in the gastrointestinal tract hydrolyze the triglycerides, releasing fatty acids and monoglycerides that facilitate the solubilization of cannabinoids into the mixed micelles [97].

The absorption of cannabinoids is strongly influenced by their lipophilic nature, and governed by a combination of passive diffusion, potential transporter interactions, lymphatic transport as well as presystemic metabolism, all of which strongly modulated by dietary lipids and formulation strategies [98]. Figure 2.4 represents the transport of cannabinoids into the human body. Their lipophilicity enables dissolution in lipid membranes, facilitation permeation across the intestinal epithelium primarily through passive diffusion across phospholipid bilayer [99]. Nonetheless, some evidence suggests that the cannabinoids permeability may also be influenced by transporters, although this remains an area of ongoing debate. According to Perucca and Bialer [100], CBD inhibits intestinal eflux transporters in the intestine but does not appear to act as a substrate. Moreover, CBD has been reported to downregulate P-glycoprotein (P-gp) expression and exert concentration-dependent inhibition of P-pg-mediated efflux, thereby enhancing intracellular accumulation of P-gp substrates [101, 102]. Overall, these absorption pathways from gastrointestinal tract into the portal vein represent the best characterized mechanism by which orally ingested bioactive compounds, including cannabinoids, enter the systemic circulation (Fig. 2.4) [103].

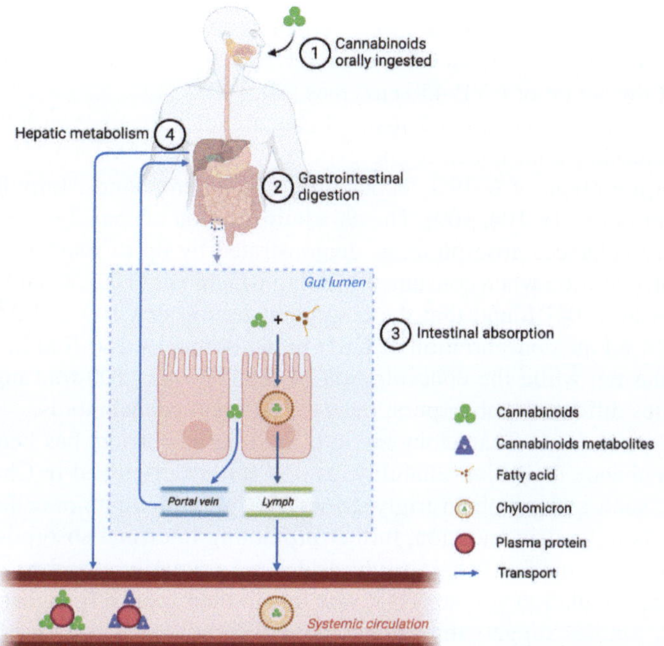

Fig. 2.4 Schematic representation of cannabinoids absorption following oral ingestion [98]. Created with BioRender.com

However, the high lipophilicity of cannabinoids also favors their absorption through the intestinal lymphatic system via incorporation into chylomicrons (Fig. 2.4), which are lipoproteins assembled in enterocytes in the presence of long-chain fatty triglycerides [104]. Although the blood flow through the portal vein is 100- to 500-fold higher than the flow rate of intestinal lymph fluid, increasing evidence indicates that absorption of lipophilic molecules through the lymphatic system may represent a major alternative route to deliver these compounds into the system circulation [103]. The lymphatic transport bypasses the portal vein and initial hepatic metabolism, thereby reducing first-pass elimination but also delaying peak plasma concentrations [105, 106]. The efficiency of lymphatic absorption increases in the presence of dietary lipids, particularly long-chain fatty acids, which improve cannabinoid incorporation into chylomicrons and promote systemic uptake [105, 106].

Once absorbed, cannabinoids undergo presystemic metabolism in both the intestinal epithelium and the liver (Fig. 2.4). Enzymes from the cytochrome P450 (CYP-450), expressed in the gastrointestinal tract, are involved in the biotransformation of cannabinoids [107, 108]. Although intestinal metabolism appears to be limited, hepatic first-pass metabolism substantially reduces systemic availability [103]. The liver extraction ratio of cannabinoids can reach 65–70%, reflecting their classification as high hepatic clearance compounds [100]. The major primary metabolites, such as 7-OH-CBD from CBD and 11-OH-THC from Δ^9-THC, are generated as result of the action of CYP-450 enzymes [88, 100].

The extensive first-pass metabolism of cannabinoids together with their as poor aqueous solubility contributes to the extremely low oral bioavailability, which is reported in the range of 6–19%, depending on the compounds, formulations, and study conditions [103, 104, 109]. The co-administration of cannabinoids with lipids significantly enhances absorption, as demonstrated by up to fourfold increases in plasma concentration when consumed with high-fat meals [100]. A study conducted by Zgair et al. [104] found that the co-administration of CBD and Δ^9-THC with lipids increased the concentration of CBD in the lymph by 250-fold in comparison with the plasma, while the concentration of Δ^9-THC was 100-fold higher, which demonstrates differential absorption rates for different cannabinoids.

The incorporation of cannabinoids into lipid-based carriers has been shown to markedly enhance their bioavailability, as it is further discussed in Chap. 5. Lipid excipients, such as long-chain triglycerides, not only facilitate micelle formation, but also prolong gastric retention, further promoting intestinal absorption [97, 105, 106]. In contrast, medium-chain triglycerides are primarily absorbed directly into the systemic circulation via the portal vein due to their relatively higher water solubility [97]. Studies suggest that cannabinoid absorption is strongly influenced by the interaction between their formulation and the composition of the food matrix, which underscores the critical role of the delivery system in determining overall bioavailability [99].

2.5.2 Distribution

The distribution of cannabinoids through the systemic circulation plays a decisive role in shaping their bioavailability, pharmacokinetics, and therapeutic potential. Understanding this process requires examining their post-absorption fate, including tissue distribution and interactions with physiological systems. In the bloodstream, cannabinoids bind extensively to plasma proteins (Fig. 2.4), which increases their apparent half-life and modulates their bioavailability [110]. Estimates indicate that around 60% of cannabinoids are bound to lipoproteins, 28% associate with albumin, and approximately 9% are linked to blood cells [97]. The lipophilic nature of cannabinoids defines their rapid distribution into the brain, adipose tissue, and other fat-rich tissues [98]. A study evaluated the pharmacokinetic and accumulation of CBD in muscle, liver, and adipose tissue in adult rats following oral administration of CBD at doses of 30, 115, and 230 mg/kg/day for 28 days [111]. The results revealed that the CBD levels in adipose tissue were approximately 10- to 100-fold greater than in muscle of liver, and this accumulation is gender specific, with a higher accumulation in female animals. The authors also reported that these findings correlated with those observed for Δ^9-THC in humans.

Binding to plasma components helps protect cannabinoids from rapid metabolic degradation and contributes to the regulation of their distribution across tissues. Moreover, association with lipoproteins can facilitate their transport and delivery to specific target sites [110]. Cannabinoid interactions with transport proteins are essential, as these compounds can act both as substrates and inhibitors of transporters, including ATP-binding cassette (ABC) proteins such as P-glycoprotein [112]. This interaction can significantly influence both the degree of cannabinoid absorption and their clearance rates.

After the initial hepatic processing, cannabinoids traverse systemic circulation, enabling them to act on various organ systems. A key target is the endocannabinoid system (ECS), which is widely distributed across the body, encompassing both the central nervous system and peripheral tissues [113]. The primary mechanism of cannabinoid action involves interaction with cannabinoid receptors (CB1 and CB2), members of the G-protein coupled receptor family found in multiple tissues, including the brain, liver, and peripheral organs. Engagement with these receptors influences diverse physiological processes such as pain perception and inflammation, highlighting the significance of their transport dynamics and interactions with the ECS [113]. Notably, they can promote satiety by modulating hormones involved in appetite control, including ghrelin and glucagon-like peptide-1 (GLP-1) and they can also influence gut-brain signaling through the vagus nerve, helping to regulate food intake and maintain energy balance in response to dietary stimuli [114, 115]. Lal et al. indicate that cannabinoid receptors, particularly CB1, can be upregulated during inflammation and gastrointestinal disorders, suggesting that cannabinoids may help mitigate symptoms of these conditions by modulating receptor activity [116].

Recent studies have emphasized the importance of fatty acid-binding proteins (FABPs) in the intracellular transport of cannabinoids. FABPs assist in shuttling hydrophobic cannabinoids through the aqueous cytosol to their target locations, including the cell membrane and various organelles [110]. The interaction between FABPs and cannabinoids highlights the importance of investigating therapeutic approaches that could target and modulate these intracellular transport pathways [113, 117].

2.5.3 Metabolism

When cannabinoids are ingested through food or dietary supplements, they undergo complex metabolic transformations involving multiple biochemical pathways and regulatory mechanisms. These processes not only determine their absorption, distribution, and elimination but also contribute to their ability to modulate feeding behavior and energy homeostasis. Understanding the metabolic fate of cannabinoids is therefore crucial, particularly for Δ^9-THC and CBD, which exert distinct by significant effects on appetite regulation, energy balance, and neuroendocrine signaling. Such effects are mediated by interactions with the ECS, an extensive signaling network involved in controlling feeding behavior and metabolic processes [98].

After oral ingestion and absorption, Δ^9-THC undergoes extensive first-pass metabolism in the liver, primarily mediated by P450 enzymes, particularly the isoforms CYP2C9, CYP2C19, and CYP3A4. This process generates multiple metabolites, the most pharmacologically relevant of which is 11-OH-THC [118, 119]. This metabolite is of particular relevance because it exhibits stronger psychoactive properties than the parent compound and crosses the blood–brain barrier more efficiently [120]. Consequently, the conversion of Δ^9-THC into 11-OH-THC substantially enhances its psychoactive, particularly following oral administration [121]. Interindividual differences in CYP2C9 activity, due to genetic variations or enzyme inhibition, can markedly influence Δ^9-THC metabolism, thereby impacting both therapeutic efficacy and the risk of adverse effects [119]. Following oxidation, Δ^9-THC metabolism undergoes conjugation through glucuronidation, during which metabolites such as 11-OH-THC and THC-COOH are linked to glucuronic acid [122]. This process is mainly mediated by uridine diphosphate-glucuronosyltransferase (UGT) enzymes, particularly UGT1A1 and UGT1A3, which enhance metabolite hydrophilicity, thereby improving solubility and facilitating excretion via urine and bile [123] .

The metabolism of CBD, while sharing some similarities with Δ^9-THC, follows distinct pathways. The primary hepatic enzymes responsible are CYP3A4 and CYP2C19, though additional enzymes such as CYP1A1, CYP1A2, CYP2C9, CYP2D6, and CYP3A5 may also contribute [107]. The process begins with hydroxylation, most commonly at the C-7 position, followed by subsequent oxidation reactions [124]. Major metabolites include 7-COOH-CBD, which is the predominant circulating form; 7-OH-CBD, an active metabolite with pharmacological activity; and 6-OH-CBD, a minor circulating metabolite [98]. CYP3A4, located in the endoplasmic reticulum of hepatocytes and also present in enterocytes, plays a critical role in the first-pass metabolism, catalyzing the formation of 7-OH-CBD, 6-OH-CBD, and additional metabolites [125]. The subsequent formation of the inactive 7-COOH-CBD involves CYP3A4, CYP2C19, and CYP2D6. Similar to Δ^9-THC, CBD metabolites undergo glucuronidation, primarily mediated by UGT1A9, UGT2A7, and UGT2B7, which produce water-soluble derivatives that are ultimately excreted in urine or feces [98].

Beyond this hepatic metabolism, cannabinoids influence metabolic and neuroendocrine processes. Through interactions with the ECS, cannabinoids can modulate hormonal responses critical for maintaining energy balance, including the regulation of insulin and leptin signaling pathways [98]. These interactions highlight the broader role of cannabinoids in the control of feeding behavior, glucose homeostasis, and lipid metabolism. Importantly, such mechanisms have raised interest in the potential therapeutic applications of cannabinoids for maintaining obesity, diabetes, and other metabolic disorders [98].

References

1. Laaboudi FZ, Rejdali M, Amhamdi H, Salhi A, Elyoussfi A, Ahari M (2024) In the weeds: a comprehensive review of cannabis; its chemical complexity, biosynthesis, and healing abilities. Toxicol Rep 13:101685. https://doi.org/10.1016/j.toxrep.2024.101685
2. Gaoni Y, Mechoulam R (1964) Isolation, structure, and partial synthesis of an active constituent of hashish. J Am Chem Soc 86:1646–1647. https://doi.org/10.1021/ja01062a046
3. Radwan MM, Chandra S, Gul S, ElSohly MA (2021) Cannabinoids, phenolics, terpenes and alkaloids of cannabis. Molecules 26:2774. https://doi.org/10.3390/molecules26092774
4. Sainz Martinez A, Lanaridi O, Stagel K, Halbwirth H, Schnürch M, Bica-Schröder K (2023) Extraction techniques for bioactive compounds of cannabis. Nat Prod Rep 40:676–717. https://doi.org/10.1039/D2NP00059H
5. dos Santos NA, Romão W (2023) Cannabis—a state of the art about the millenary plant: part I. Forensic Chem 32:100470. https://doi.org/10.1016/j.forc.2023.100470
6. Duggan PJ (2021) The chemistry of cannabis and cannabinoids. Aust J Chem 74:369–387. https://doi.org/10.1071/CH21006
7. Blebea NM, Hancu G, Vlad RA, Pricopie A (2023) Applications of capillary electrophoresis for the determination of cannabinoids in different matrices. Molecules 28:638. https://doi.org/10.3390/molecules28020638
8. Addo PW, Desaulniers Brousseau V, Morello V, MacPherson S, Paris M, Lefsrud M (2021) Cannabis chemistry, post-harvest processing methods and secondary metabolite profiling: a review. Ind Crops Prod 170:113743. https://doi.org/10.1016/j.indcrop.2021.113743
9. Ueda N, Tsuboi K, Uyama T (2013) Metabolism of endocannabinoids and related <scp>N</scp>-acylethanolamines: canonical and alternative pathways. FEBS J 280:1874–1894. https://doi.org/10.1111/febs.12152
10. Shevyrin VA, Morzherin YuYu (2015) Cannabinoids: structures, effects, and classification. Russ Chem Bull 64:1249–1266. https://doi.org/10.1007/s11172-015-1008-1
11. Teixeira HM (2025) Phytocanabinoids and synthetic cannabinoids: from recreational consumption to potential therapeutic use—a review. Front Toxicol 6. https://doi.org/10.3389/ftox.2024.1495547
12. Cascio MG (2013) PUFA-derived endocannabinoids: an overview. Proceed Nutr Soc 72:451–459. https://doi.org/10.1017/S0029665113003418
13. Barutta F, Bruno G, Mastrocola R, Bellini S, Gruden G (2018) The role of cannabinoid signaling in acute and chronic kidney diseases. Kidney Int 94:252–258. https://doi.org/10.1016/J.KINT.2018.01.024
14. Francois H, Lecru L (2018) The role of cannabinoid receptors in renal diseases. Curr Med Chem 25:793–801. https://doi.org/10.2174/0929867324666170911170020/CITE/REF WORKS
15. Freitas HR, Isaac AR, Malcher-Lopes R, Diaz BL, Trevenzoli IH, De Melo Reis RA (2018) Polyunsaturated fatty acids and endocannabinoids in health and disease. Nutr Neurosci 21:695–714. https://doi.org/10.1080/1028415X.2017.1347373
16. Pete DD, Narouze SN (2021) Endocannabinoids: anandamide and 2-arachidonoylglycerol (2-AG). Cannabinoids Pain 63–69. https://doi.org/10.1007/978-3-030-69186-8_9
17. Blankman JL, Cravatt BF (2013) Chemical probes of endocannabinoid metabolism. Pharmacol Rev 65:849–871. https://doi.org/10.1124/pr.112.006387
18. Nomura DK, Morrison BE, Blankman JL, Long JZ, Kinsey SG, Marcondes MCG, Ward AM, Hahn YK, Lichtman AH, Conti B, Cravatt BF (1979) Endocannabinoid hydrolysis generates brain prostaglandins that promote neuroinflammation. Science 334(2011):809–813. https://doi.org/10.1126/science.1209200
19. Lu H-C, Mackie K (2016) An introduction to the endogenous cannabinoid system. Biol Psychiatry 79:516–525. https://doi.org/10.1016/j.biopsych.2015.07.028
20. Alves VL, Gonçalves JL, Aguiar J, Teixeira HM, Câmara JS (2020) The synthetic cannabinoids phenomenon: from structure to toxicological properties. A Rev Crit Rev Toxicol 50:359–382. https://doi.org/10.1080/10408444.2020.1762539

21. Alghamdi SS, Albahlal HN, Aloumi DE, Bin Saqyah S, Alsubait A, Alamre J, Alrashed M, Alsuhabeny N, Mohammed AE (2025) Revealing the therapeutic potential of synthetic cannabinoids: a systematic review of cannabinoid receptor binding dynamics and their implications for cancer therapy. J Cannabis Res 7:33. https://doi.org/10.1186/s42238-025-00289-5

22. Maglaviceanu A, Peer M, Rockel J, Bonin RP, Fitzcharles M-A, Ladha KS, Bhatia A, Leroux T, Kotra L, Kapoor M, Clarke H (2024) The state of synthetic cannabinoid medications for the treatment of pain. CNS Drugs 38:597–612. https://doi.org/10.1007/s40263-024-01098-9

23. de Oliveira MC, Vides MC, Lassi DLS, Torales J, Ventriglio A, Bombana HS, Leyton V, Périco DAM, Negrão AB, Malbergier A, Castaldelli-Maia JM (2023) Toxicity of synthetic cannabinoids in K2/spice: a systematic review. Brain Sci 13: 990. https://doi.org/10.3390/brainsci13070990

24. Lafaye G, Karila L, Blecha L, Benyamina A (2017) Cannabis, cannabinoids, and health. Dialogues Clin Neurosci 19:309–316. https://doi.org/10.31887/DCNS.2017.19.3/glafaye

25. Ottani A, Giuliani D (2001) HU 210: a potent tool for investigations of the cannabinoid system. CNS Drug Rev 7:131–145. https://doi.org/10.1111/j.1527-3458.2001.tb00192.x

26. Cohen K, Weinstein A (2018) The effects of cannabinoids on executive functions: evidence from cannabis and synthetic cannabinoids—a systematic review. Brain Sci 8:40. https://doi.org/10.3390/brainsci8030040

27. Huffman JW, Szklennik PV, Almond A, Bushell K, Selley DE, He H, Cassidy MP, Wiley JL, Martin BR (2005) 1-Pentyl-3-phenylacetylindoles, a new class of cannabimimetic indoles. Bioorg Med Chem Lett 15:4110–4113. https://doi.org/10.1016/j.bmcl.2005.06.008

28. Brents LK, Prather PL (2014) The K2/Spice Phenomenon: emergence, identification, legislation and metabolic characterization of synthetic cannabinoids in herbal incense products. Drug Metab Rev 46:72–85. https://doi.org/10.3109/03602532.2013.839700

29. Cooper ZD (2016) Adverse effects of synthetic cannabinoids: management of acute toxicity and withdrawal. Curr Psychiatry Rep 18:52. https://doi.org/10.1007/s11920-016-0694-1

30. Govindarajan RK, Mishra AK, Cho K-H, Kim K-H, Yoon KM, Baek K-H (2023) Biosynthesis of phytocannabinoids and structural insights: a review. Metabolites 13:442. https://doi.org/10.3390/metabo13030442

31. Yadav SPS, Kafle M, Ghimire NP, Shah NK, Dahal P, Pokhrel S (2023) An overview of phytochemical constituents and pharmacological implications of Cannabis sativa L. J Herb Med 42:100798. https://doi.org/10.1016/j.hermed.2023.100798

32. Thorsen TS, Kulkarni Y, Sykes DA, Bøggild A, Drace T, Hompluem P, Iliopoulos-Tsoutsouvas C, Nikas SP, Daver H, Makriyannis A, Nissen P, Gajhede M, Veprintsev DB, Boesen T, Kastrup JS, Gloriam DE (2025) Structural basis of THC analog activity at the Cannabinoid 1 receptor. Nat Commun 16:486. https://doi.org/10.1038/s41467-024-55808-4

33. Lazarjani MP, Young O, Kebede L, Seyfoddin A (2021) Processing and extraction methods of medicinal cannabis: a narrative review. J Cannabis Res 3:1–15. https://doi.org/10.1186/S42238-021-00087-9/FIGURES/6

34. Tambunan AH, Yudistira, Kisdiyani, Hernani (2001) Freeze drying characteristics of medicinal herbs. Drying Technol 19:325–331. https://doi.org/10.1081/DRT-100102907;PAGE:STRING:ARTICLE/CHAPTER

35. Chen C, Wongso I, Putnam D, Khir R, Pan Z (2021) Effect of hot air and infrared drying on the retention of cannabidiol and terpenes in industrial hemp (Cannabis sativa L.). Ind Crops Prod 172. https://doi.org/10.1016/J.INDCROP.2021.114051

36. Hanuš LO, Meyer SM, Muñoz E, Taglialatela-Scafati O, Appendino G (2016) Phytocannabinoids: a unified critical inventory. Nat Prod Rep 33:1357–1392. https://doi.org/10.1039/C6NP00074F

37. Tanney CAS, Backer R, Geitmann A, Smith DL (2021) Cannabis glandular trichomes: a cellular metabolite factory. Front Plant Sci 12. https://doi.org/10.3389/fpls.2021.721986

38. Aizpurua-Olaizola O, Omar J, Navarro P, Olivares M, Etxebarria N, Usobiaga A (2014) Identification and quantification of cannabinoids in Cannabis sativa L. plants by high performance liquid chromatography-mass spectrometry. Anal Bioanal Chem 406:7549–7560. https://doi.org/10.1007/s00216-014-8177-x

39. Vollner L, Bieniek D, Korte F (1969) Haschisch XX. Tetrahedron Lett 10:145–147. https://doi.org/10.1016/S0040-4039(01)87494-3

40. Yamauchi T, Shoyama Y, Aramaki H, Azuma T, Nishioka I (1967) Tetrahydrocannabinolic acid, a genuine substance of tetrahydrocannabinol. Chem Pharm Bull (Tokyo) 15:1075–1076. https://doi.org/10.1248/cpb.15.1075

41. Mechoulam R, Ben-Zvi Z, Yagnitinsky B, Shani A (1969) A new tetrahydrocannabinolic acid. Tetrahedron Lett 10:2339–2341. https://doi.org/10.1016/S0040-4039(01)88158-2

42. Harvey DI (1976) Characterization of the butyl homologues of Δ1-tetrahydrocannabinol, cannabinol and cannabidiol in samples of cannabis by combined gas chromatography and mass spectrometry. J Pharm Pharmacol 28:280–285. https://doi.org/10.1111/j.2042-7158.1976.tb04153.x

43. Shoyama Y, Hirano H, Makino H, Umekita N, Nishioka I (1977) Cannabis. X. the isolation and structures of four new propyl cannabinoid acids, tetrahydrocannabivarinic acid, cannabidivarinic acid, cannabichromevarinic acid and cannabigerovarinic acid, from Thai Cannabis, "Meao variant". Chem Pharm Bull (Tokyo) 25:2306–2311. https://doi.org/10.1248/cpb.25.2306

44. Ahmed SA, Ross SA, Slade D, Radwan MM, Khan IA, ElSohly MA (2015) Minor oxygenated cannabinoids from high potency Cannabis sativa L. Phytochemistry 117:194–199. https://doi.org/10.1016/j.phytochem.2015.04.007

45. Ahmed SA, Ross SA, Slade D, Radwan MM, Zulfiqar F, ElSohly MA (2008) Cannabinoid ester constituents from high-potency *Cannabis sativa*. J Nat Prod 71:536–542. https://doi.org/10.1021/np070454a

46. Radwan M, Ross S, Slade D, Ahmed S, Zulfiqar F, ElSohly M (2008) Isolation and characterization of new cannabis constituents from a high potency variety. Planta Med 74:267–272. https://doi.org/10.1055/s-2008-1034311

47. Zulfiqar F, Ross SA, Slade D, Ahmed SA, Radwan MM, Ali Z, Khan IA, ElSohly MA (2012) Cannabisol, a novel Δ9-THC dimer possessing a unique methylene bridge, isolated from Cannabis sativa. Tetrahedron Lett 53:3560–3562. https://doi.org/10.1016/j.tetlet.2012.04.139

48. Shoyama Y, Kuboe K, Nishioka I, Yamauchi T (1972) Cannabidiol Monomethyl ether. A new neutral cannabinoid. Chem Pharm Bull (Tokyo) 20:2072–2072. https://doi.org/10.1248/cpb.20.2072

49. Pattnaik F, Nanda S, Mohanty S, Dalai AK, Kumar V, Ponnusamy SK, Naik S (2022) Cannabis: Chemistry, extraction and therapeutic applications. Chemosphere 289:133012. https://doi.org/10.1016/j.chemosphere.2021.133012

50. Radwan MM, ElSohly MA, Slade D, Ahmed SA, Khan IA, Ross SA (2009) Biologically active cannabinoids from high-potency *Cannabis sativa*. J Nat Prod 72:906–911. https://doi.org/10.1021/np900067k

51. Bercht CAL, Lousberg RJJ, Küppers FJEM, Salemink CA, Vree TB, Van Rossum JM (1973) Cannabis. J Chromatogr A 81:163–166. https://doi.org/10.1016/S0021-9673(01)82332-3

52. Uliss DB, Razdan RK, Dalzell HC (1974) Stereospecific intramolecular epoxide cleavage by phenolate anion. Synthesis of novel and biologically active cannabinoids. J Am Chem Soc 96:7372–7374. https://doi.org/10.1021/ja00830a045

53. Grote H, Spiteller G (1978) Neue cannabinoide—III. Tetrahedron 34:3207–3213. https://doi.org/10.1016/0040-4020(78)87018-5

54. Korte F, Sieper H (1964) Zur chemischen klassifizierung von pflanzen. J Chromatogr A 13:90–98. https://doi.org/10.1016/S0021-9673(01)95077-0

55. Begley MJ, Clarke DG, Crombie L, Whiting DA (1970) The X-ray structure of dibromo-cannabicyclol: structure of bicyclomahanimbine. J Chem Soc D: Chem Commun 1547. https://doi.org/10.1039/c29700001547

56. Shoyama Y, Oku R, Yamauchi T, Nishioka I (1972) Cannabis. VI. cannabicyclolic acid. Chem Pharm Bull (Tokyo) 20:1927–1930. https://doi.org/10.1248/cpb.20.1927

57. Gaoni Y, Mechoulam R (1966) Cannabichromene, a new active principle in hashish. Chem Commun (Lond) 20. https://doi.org/10.1039/c19660000020

58. Shoyama Y, Yagi M, Nishioka I, Yamauchi T (1975) Biosynthesis of cannabinoid acids. Phytochemistry 14:2189–2192. https://doi.org/10.1016/S0031-9422(00)91096-3

59. Cascio MG, Pertwee RG, Marini P (2017) The pharmacology and therapeutic potential of plant cannabinoids. In: Cannabis Sativa L.—botany and biotechnology. Springer International Publishing, Cham, pp207–225. https://doi.org/10.1007/978-3-319-54564-6_9

60. Obata Y, Ishikawa Y (1966) Studies on the constituents of hemp plant (*Cannabis sativa* L.). Agric Biol Chem 30:619–620. https://doi.org/10.1080/00021369.1966.10858651

61. Taura F, Morimoto S, Shoyama Y (1995) Cannabinerolic acid, a cannabinoid from Cannabis sativa. Phytochemistry 39:457–458. https://doi.org/10.1016/0031-9422(94)00887-Y

62. Appendino G, Giana A, Gibbons S, Maffei M, Gnavi G, Grassi G, Sterner O (2008) A polar cannabinoid from *Cannabis Sativa* Var. *Carma*. Nat Prod Commun 3. https://doi.org/10.1177/1934578X0800301207

63. Chan WR, Magnus KE, Watson HA (1976) The structure of cannabitriol. Experientia 32:283–284. https://doi.org/10.1007/BF01940792

64. Elsohly MA, El-Feraly FS, Turner CE (1977) Isolation and characterization of (+)-cannabitriol and (-)-10-ethoxy-9-hydroxy-delta 6a[10a]-tetrahydrocannabinol: two new cannabinoids from Cannabis sativa L. extract. Lloydia 40:275–280

65. Russo EB, Jiang H-E, Li X, Sutton A, Carboni A, del Bianco F, Mandolino G, Potter DJ, Zhao Y-X, Bera S, Zhang Y-B, Lü E-G, Ferguson DK, Hueber F, Zhao L-C, Liu C-J, Wang Y-F, Li C-S (2008) Phytochemical and genetic analyses of ancient cannabis from Central Asia. J Exp Bot 59:4171–4182. https://doi.org/10.1093/jxb/ern260

66. Robert L, Ch JJ, Ludwig Bercht CA, van Ooyen R, Spronck HJW (1977) Cannabinodiol: conclusive identification and synthesis of a new cannabinoid from Cannabis sativa. Phytochemistry 16:595–597. https://doi.org/10.1016/0031-9422(77)80023-X

67. Hively RL, Mosher WA, Hoffmann FW (1966) Isolation of trans-Δ^6-tetrahydrocannabinol from marijuana. J Am Chem Soc 88:1832–1833. https://doi.org/10.1021/ja00960a056

68. Hanus L, Krejci Z (1975) Isolation of two new cannabinoid acids from Cannabis sativa L. of Czechoslovak origin. Acta Univ Palacki Olomuc Fac Med 74

69. Desaulniers Brousseau V, Wu BS, MacPherson S, Morello V, Lefsrud M (2021) Cannabinoids and terpenes: how production of photo-protectants can be manipulated to enhance Cannabis sativa L. Phytochem, Front Plant Sci 12. https://doi.org/10.3389/fpls.2021.620021

70. Bali S, Mohapatra S, Michael R, Arora R, Dogra V (2025) Plastidial metabolites and retrograde signaling: a case study of MEP pathway intermediate MEcPP that orchestrates plant growth and stress responses. Plant Physiol Biochem 222:109747. https://doi.org/10.1016/j.plaphy.2025.109747

71. Tahir MN, Raz FS, Rondeau-Gagné S, Trant JF (2021) The biosynthesis of the cannabinoids. J Cannabis Res 3:7. https://doi.org/10.1186/s42238-021-00062-4

72. Shahbazi-Raz F, Meister D, Mohammadzadeh A, Trant JF (2025) How THC works: explaining ligand affinity for, and partial agonism of, cannabinoid receptor 1. IScience 28:112706. https://doi.org/10.1016/j.isci.2025.112706

73. Tambe SM, Mali S, Amin PD, Oliveira M (2023) Neuroprotective potential of cannabidiol: molecular mechanisms and clinical implications. J Integr Med 21:236–244. https://doi.org/10.1016/j.joim.2023.03.004

74. Li S, Li W, Malhi NK, Huang J, Li Q, Zhou Z, Wang R, Peng J, Yin T, Wang H (2024) Cannabigerol (CBG): a comprehensive review of its molecular mechanisms and therapeutic potential. Molecules 29:5471. https://doi.org/10.3390/molecules29225471

75. Hong M, Kim J-H, Han J-H, Ryu B-R, Lim Y-S, Lim J-D, Park S-H, Kim C-H, Lee S-U, Kwon T-H (2023) In vitro and in vivo anti-inflammatory potential of cannabichromene isolated from hemp. Plants 12:3966. https://doi.org/10.3390/plants12233966

76. Lavender I, McCartney D, Marshall N, Suraev A, Irwin C, D'Rozario AL, Gordon CJ, Saini B, Grunstein RR, Yee B, McGregor I, Hoyos CM (2023) Cannabinol (CBN; 30 and 300 mg) effects on sleep and next-day function in insomnia disorder ('CUPID' study): protocol for a randomised, double-blind, placebo-controlled, cross-over, three-arm, proof-of-concept trial. BMJ Open 13:e071148. https://doi.org/10.1136/bmjopen-2022-071148

77. Banerjee A, Hayward JJ, Trant JF (2023) "Breaking bud": the effect of direct chemical modifications of phytocannabinoids on their bioavailability, physiological effects, and therapeutic potential. Org Biomol Chem 21:3715–3732. https://doi.org/10.1039/D3OB00068K

78. Kesavan Pillai S, Hassan Kera N, Kleyi P, de Beer M, Magwaza M, Ray SS (2024) Stability, biofunctional, and antimicrobial characteristics of cannabidiol isolate for the design of topical formulations. Soft Matter 20:2348–2360. https://doi.org/10.1039/D3SM01466E

79. Seo C, Jeong M, Lee S, Kim EJ, Rho S, Cho M, Lee YS, Hong J (2022) Thermal decarboxylation of acidic cannabinoids in Cannabis species: identification of transformed cannabinoids by UHPLC-Q/TOF–MS. J Anal Sci Technol 13:42. https://doi.org/10.1186/s40543-022-003 51-4

80. Park C, Zuo J, Somayaji V, Lee B-J, Löbenberg R (2021) Development of a novel cannabinoid-loaded microemulsion towards an improved stability and transdermal delivery. Int J Pharm 604:120766. https://doi.org/10.1016/j.ijpharm.2021.120766

81. Eyal AM, Berneman Zeitouni D, Tal D, Schlesinger D, Davidson EM, Raz N (2023) Vapor pressure, vaping, and corrections to misconceptions related to medical cannabis' active pharmaceutical ingredients' physical properties and compositions. Cannabis Cannabinoid Res 8:414–425. https://doi.org/10.1089/can.2021.0173

82. McPartland JM, Russo EB (2001) Cannabis and cannabis extracts. J Cannabis Ther 1:103–132. https://doi.org/10.1300/J175v01n03_08

83. Mathew M, Knapik-Kowalczuk J, Dulski M, Paluch M (2025) High-pressure dielectric spectroscopic studies of amorphous CBD: investigating molecular dynamics and physical stability under manufacturing conditions of the pharmaceuticals. Pharmaceutics 17:358. https://doi.org/10.3390/pharmaceutics17030358

84. Filer CN (2022) Cannabinoid crystal polymorphism. J Cannabis Res 4:23. https://doi.org/10.1186/s42238-022-00131-2

85. Taha IE, ElSohly MA, Radwan MM, Elkanayati RM, Wanas A, Joshi PH, Ashour EA (2025) Enhancement of cannabidiol oral bioavailability through the development of nanostructured lipid carriers: in vitro and in vivo evaluation studies. Drug Deliv Transl Res 15:2722–2732. https://doi.org/10.1007/s13346-024-01766-9

86. Chayasirisobhon S (2021) Mechanisms of action and pharmacokinetics of cannabis. Perm J 25:1–3. https://doi.org/10.7812/TPP/19.200

87. Mahadevan A, Siegel C, Martin BR, Abood ME, Beletskaya I, Razdan RK (2000) Novel cannabinol probes for CB1 and CB2 cannabinoid receptors. J Med Chem 43:3778–3785. https://doi.org/10.1021/jm0001572

88. Rao Q, Zhang T, Dai M, Li B, Pu Q, Zhao M, Cheng Y, Yan D, Zhao Q, Wu ZE, Li F (2022) Comparative metabolomic profiling of the metabolic differences of $\Delta 9$-tetrahydrocannabinol and cannabidiol. Molecules 27:7573. https://doi.org/10.3390/molecules27217573

89. Docampo-Palacios ML, Ramirez GA, Tesfatsion TT, Okhovat A, Pittiglio M, Ray KP, Cruces W (2023) Saturated cannabinoids: update on synthesis strategies and biological studies of these emerging cannabinoid analogs. Molecules 28:6434. https://doi.org/10.3390/MOLECU LES28176434

90. Nelson KM, Bisson J, Singh G, Graham JG, Chen S-N, Friesen JB, Dahlin JL, Niemitz M, Walters MA, Pauli GF (2020) The essential medicinal chemistry of cannabidiol (CBD). J Med Chem 63:12137–12155. https://doi.org/10.1021/acs.jmedchem.0c00724

91. Bow EW, Rimoldi JM (2016) The structure-function relationships of classical cannabinoids: CB1/CB2 modulation. Perspect Med Chem 8:17. https://doi.org/10.4137/PMC.S32171

92. Shahbazi F, Grandi V, Banerjee A, Trant JF (2020) Cannabinoids and cannabinoid receptors: the story so far. IScience 23:101301. https://doi.org/10.1016/J.ISCI.2020.101301

93. Tagen M, Klumpers LE (2022) Review of delta-8-tetrahydrocannabinol ($\Delta 8$-THC): comparative pharmacology with $\Delta 9$-THC. Br J Pharmacol 179:3915–3933. https://doi.org/10.1111/BPH.15865

94. Henriques A (2025) Cannabinoid spoilage, metabolism and cannabidiol(CBD) conversion to tetrahydrocannabinol(THC) mechanisms with energetic parameters. J Cannabis Res 7:11. https://doi.org/10.1186/s42238-024-00239-7

95. Teixeira HM (2024) Phytocanabinoids and synthetic cannabinoids: from recreational consumption to potential therapeutic use—a review. Front Toxicol 6:1495547. https://doi.org/10.3389/FTOX.2024.1495547/BIBTEX

96. Abdollahzadeh Hamzekalayi MR, Hooshyari Ardakani M, Moeini Z, Rezaei R, Hamidi N, Rezaei Somee L, Zolfaghar M, Darzi R, Kamalipourazad M, Riazi G, Meknatkhah S (2024) A systematic review of novel cannabinoids and their targets: Insights into the significance of structure in activity. Eur J Pharmacol 976:176679. https://doi.org/10.1016/j.ejphar.2024.176679

97. McClements DJ (2020) Enhancing efficacy, performance, and reliability of cannabis edibles: insights from lipid bioavailability studies. Annu Rev Food Sci Technol 45–70. https://doi.org/10.1146/annurev-food-032519

98. Sitovs A, Logviss K, Lauberte L, Mohylyuk V (2024) Oral delivery of cannabidiol: revealing the formulation and absorption challenges. J Drug Deliv Sci Technol 92:105316. https://doi.org/10.1016/J.JDDST.2023.105316

99. Tabboon P, Pongjanyakul T, Limpongsa E, Jaipakdee N (2022) In vitro release, mucosal permeation and deposition of cannabidiol from liquisolid systems: the influence of liquid vehicles. Pharmaceutics 14:1787. https://doi.org/10.3390/pharmaceutics14091787

100. Perucca E, Bialer M (2020) Critical aspects affecting cannabidiol oral bioavailability and metabolic elimination, and related clinical implications. CNS Drugs 34:795–800. https://doi.org/10.1007/s40263-020-00741-5

101. Zhu HJ, Wang JS, Markowitz JS, Donovan JL, Gibson BB, Gefroh HA, DeVane CL (2006) Characterization of p-glycoprotein inhibition by major cannabinoids from marijuana. J Pharmacol Exp Ther 317:850–857. https://doi.org/10.1124/JPET.105.098541

102. Holland ML, Panetta JA, Hoskins JM, Bebawy M, Roufogalis BD, Allen JD, Arnold JC (2006) The effects of cannabinoids on P-glycoprotein transport and expression in multidrug resistant cells. Biochem Pharmacol 71:1146–1154. https://doi.org/10.1016/J.BCP.2005.12.033

103. Franco V, Gershkovich P, Perucca E, Bialer M (2020) The interplay between liver first-pass effect and lymphatic absorption of cannabidiol and its implications for cannabidiol oral formulations. Clin Pharmacokinet 59:1493–1500. https://doi.org/10.1007/s40262-020-00931-w

104. Zgair A, Lee JB, Wong JCM, Taha DA, Aram J, Di Virgilio D, McArthur JW, Cheng Y-K, Hennig IM, Barrett DA, Fischer PM, Constantinescu CS, Gershkovich P (2017) Oral administration of cannabis with lipids leads to high levels of cannabinoids in the intestinal lymphatic system and prominent immunomodulation. Sci Rep 7:14542. https://doi.org/10.1038/s41598-017-15026-z

105. Izgelov D, Shmoeli E, Domb AJ, Hoffman A (2020) The effect of medium chain and long chain triglycerides incorporated in self-nano emulsifying drug delivery systems on oral absorption of cannabinoids in rats. Int J Pharm 580. https://doi.org/10.1016/j.ijpharm.2020.119201

106. De Prá MAA, Vardanega R, Loss CG (2021) Lipid-based formulations to increase cannabidiol bioavailability: in vitro digestion tests, pre-clinical assessment and clinical trial. Int J Pharm 609 (2021). https://doi.org/10.1016/j.ijpharm.2021.121159

107. Stout SM, Cimino NM (2014) Exogenous cannabinoids as substrates, inhibitors, and inducers of human drug metabolizing enzymes: a systematic review. Drug Metab Rev 46:86–95. https://doi.org/10.3109/03602532.2013.849268

108. Anzenbacher P, Anzenbacherová E (2001) Cytochromes P450 and metabolism of xenobiotics. Cell Mol Life Sci 58:737–747. https://doi.org/10.1007/PL00000897

109. Scientific Opinion on the risks for human health related to the presence of tetrahydro-cannabinol (THC) in milk and other food of animal origin. EFSA J 13. https://doi.org/10.2903/j.efsa.2015.4141

110. Elmes MW, Prentis LE, McGoldrick LL, Giuliano CJ, Sweeney JM, Joseph OM, Che J, Carbonetti GS, Studholme K, Deutsch DG, Rizzo RC, Glynn SE, Kaczocha M (2019) FABP1 controls hepatic transport and biotransformation of Δ9-THC. Sci Rep 9:7588. https://doi.org/10.1038/s41598-019-44108-3

111. Child RB, Tallon MJ (2022) Cannabidiol (CBD) dosing: plasma pharmacokinetics and effects on accumulation in skeletal muscle, liver and adipose tissue. Nutrients 14:2101. https://doi.org/10.3390/nu14102101

112. Anderson LL, Etchart MG, Bahceci D, Golembiewski TA, Arnold JC (2021) Cannabis constituents interact at the drug efflux pump BCRP to markedly increase plasma cannabidiolic acid concentrations. Sci Rep 11:14948. https://doi.org/10.1038/s41598-021-94212-6

113. Rosado-Franco J, Ellison A, White C, Price A, Moore C, Williams R, Fridman L, Weerts E, Williams D (2023). Roadmap for the expression of canonical and extended endocannabinoid system receptors and proteins in peripheral organs of preclinical animal models. https://doi.org/10.1101/2023.06.10.544455

114. Sharkey KA, Wiley JW (2016) The role of the endocannabinoid system in the brain-gut axis. Gastroenterology 151:252–266. https://doi.org/10.1053/j.gastro.2016.04.015

115. Capasso R, Izzo AA (2008) Gastrointestinal regulation of food intake: general aspects and focus on anandamide and oleoylethanolamide. J Neuroendocrinol 20:39–46. https://doi.org/10.1111/j.1365-2826.2008.01686.x

116. Lal S, Prasad N, Ryan M, Tangri S, Silverberg MS, Gordon A, Steinhart H (2011) Cannabis use amongst patients with inflammatory bowel disease. Eur J Gastroenterol Hepatol 23:891–896. https://doi.org/10.1097/MEG.0b013e328349bb4c

117. Lin Y-F (2021) Potassium channels as molecular targets of endocannabinoids. Channels 15:408–423. https://doi.org/10.1080/19336950.2021.1910461

118. Antoniou T, Bodkin J, Ho JM-W (2020) Drug interactions with cannabinoids. Can Med Assoc J 192:E206–E206. https://doi.org/10.1503/cmaj.191097

119. Nachnani R, Knehans A, Neighbors JD, Kocis PT, Lee T, Tegeler K, Trite T, Raup-Konsavage WM, Vrana KE (2024) Systematic review of drug-drug interactions of delta-9-tetrahydrocannabinol, cannabidiol, and cannabis. Front Pharmacol 15. https://doi.org/10.3389/fphar.2024.1282831

120. Leghissa A, Smuts J, Qiu C, Hildenbrand ZL, Schug KA (2018) Detection of cannabinoids and cannabinoid metabolites using gas chromatography with vacuum ultraviolet spectroscopy. Sep Sci Plus 1:37–42. https://doi.org/10.1002/sscp.201700005

121. Peng H, Shahidi F (2021) Cannabis and cannabis edibles: a review. J Agric Food Chem 69:1751–1774. https://doi.org/10.1021/acs.jafc.0c07472

122. Nasrin S, Watson CJW, Bardhi K, Fort G, Chen G, Lazarus P (2021) Inhibition of UDP-glucuronosyltransferase enzymes by major cannabinoids and their metabolites. Drug Metab Dispos 49:1081–1089. https://doi.org/10.1124/dmd.121.000530

123. Davies M, Peramuhendige P, King L, Golding M, Kotian A, Penney M, Shah S, Manevski N (2020) Evaluation of in vitro models for assessment of human intestinal metabolism in drug discovery. Drug Metab Dispos 48:1169–1182. https://doi.org/10.1124/dmd.120.000111

124. Mechoulam R, Hanuš L (2002) Cannabidiol: an overview of some chemical and pharmacological aspects. Part I: chemical aspects. Chem Phys Lipids 121:35–43. https://doi.org/10.1016/S0009-3084(02)00144-5

125. Zhang Y, Benet LZ (2001) The gut as a barrier to drug absorption. Clin Pharmacokinet 40:159–168. https://doi.org/10.2165/00003088-200140030-00002

Chapter 3
Extraction of Cannabinoids

There is a growing interest in the use of cannabis-based products across various industrial sectors, particularly those containing cannabinoids with proven biological activity [1]. Typically, the development of products that take advantage of the bioactive and technological potential of cannabinoids relies on solid–liquid extraction processes, which is a crucial step in the production chain [2]. Figure 3.1 illustrates a manufacturing process currently employed for the production and purification of cannabis extracts for subsequent applications. This process is complex and involves multiple steps, with the extraction technique playing a critical role in determining the characteristics of the final product. Several technologies have been explored to separate the cannabinoids from the plant material, including conventional techniques such as solvent extraction and mechanical pressing, as well as more advanced methods, including ultrasound-assisted extraction (UAE), microwave-assisted extraction (MAE), pressurized liquid extraction (PLE), supercritical fluid extraction (SFE), ionic liquids and deep eutectic solvents, and enzymatic-assisted extraction (EAE) [2–4]. The main techniques at advanced stage of development for the extraction of cannabinoids, namely, solvent extraction, UAE, PLE, and SFE are discussed in this section.

As illustrated in Fig. 3.1, the manufacturing process to produce cannabis extracts starts with the grinding of the raw material, frequently followed by decarboxylation to activate the cannabinoids. The decarboxylated material is then subjected to the extraction step to recover the target compounds. The choice of the extraction technique determines subsequent purification steps. For instance, when supercritical CO_2 is employed in the supercritical fluid extraction (SFE), the resulting extract is typically subjected to winterization, via precipitation in cold ethanol, to separate the co-extracted waxes. This is followed by ethanol removal and further concentration of the cannabinoids by vacuum distillation. In contrast, when extraction techniques involving ethanol as the primary solvent are used, winterization is not necessary. The terpenes lost during the processing can be added back to the final distillate, which is often incorporated into the food matrices. Alternatively, further purification of

© The Author(s), under exclusive license to Springer Nature Switzerland AG 2026   35
S. E. Azam et al., *Cannabinoids in Food Science*,
Chemistry of Foods, https://doi.org/10.1007/978-3-032-11546-1_3

the distillate can be achieved by crystallization techniques to obtain cannabinoids isolate.

3.1 Solvent Extraction

Conventionally, solvent extraction refers to methods such as Soxhlet and maceration. In Soxhlet extraction, distillation is the main mechanism involved. In this method, the sample is placed in a porous container within the extraction chamber. The solvent is heated in a distillation flask, evaporates and condensates into the chamber gradually filling it with fresh solvent. Once the solvent reaches a set level, it is siphoned back into the boiling flask, carrying with it the extracted compounds. This cycle is continuously repeated, allowing continuous contact between sample and clean solvent, enabling efficient and exhaustive extraction of compounds [2, 5]. Soxhlet extraction is extensively applied to recover bioactive compounds from plant matter and considered as a reference method for development of new alternative methods of separation, despite it requires large amounts of solvent and is time- and energy-intensive [6].

Maceration is also frequently used to obtain essential oils and bioactive compounds from plant materials. This method is based on the diffusion of plant compounds into the organic solvents and consists of soaking the plant material in organic solvents for a defined time at a specific temperature followed by agitation to increase the diffusion of the target compounds [6]. The efficiency of this method relies primarily on the polarity of the target compounds, as well as the diffusion mechanisms across plant cell walls [7]. In maceration, agitation plays an important role in enhancing mass transfer by promoting better contact between the solvent and plant material, thus increasing the extraction efficiency [6]. Although simple and inexpensive, maceration typically requires long extraction times and high amounts of solvent, which limits its scalability and sustainability. Nonetheless, maceration remains popular in homemade applications to recover essential oils and other bioactive compounds [2].

The efficiency of these conventional methods strongly relies on the solubility of the target compounds in the chosen solvents. Cannabinoids are relatively non-polar due to the presence of a predominant proportion of carbon and hydrogen, which makes them preferably well dissolved in organic solvents [4]. The most preferred organic solvents for cannabinoids extraction include ethanol, butane, hexane, methanol, and acetone [2], but mixtures of solvents are sometimes used to improve the extraction efficiency [4]. Ethanol is usually the most preferred solvent for cannabinoids extraction. However, its low selectivity promotes the extraction of unwanted compounds, such as chlorophyll, which adds an unpleasant flavor and a green color to the end product, thus requiring additional refining processes [2].

Recently, vegetable oils have been considered as an interesting alternative solvent to recover cannabinoids due to their non-polar characteristic, which enables selectively dissolving lipophilic compounds, such as terpenes and cannabinoids [8]. Olive

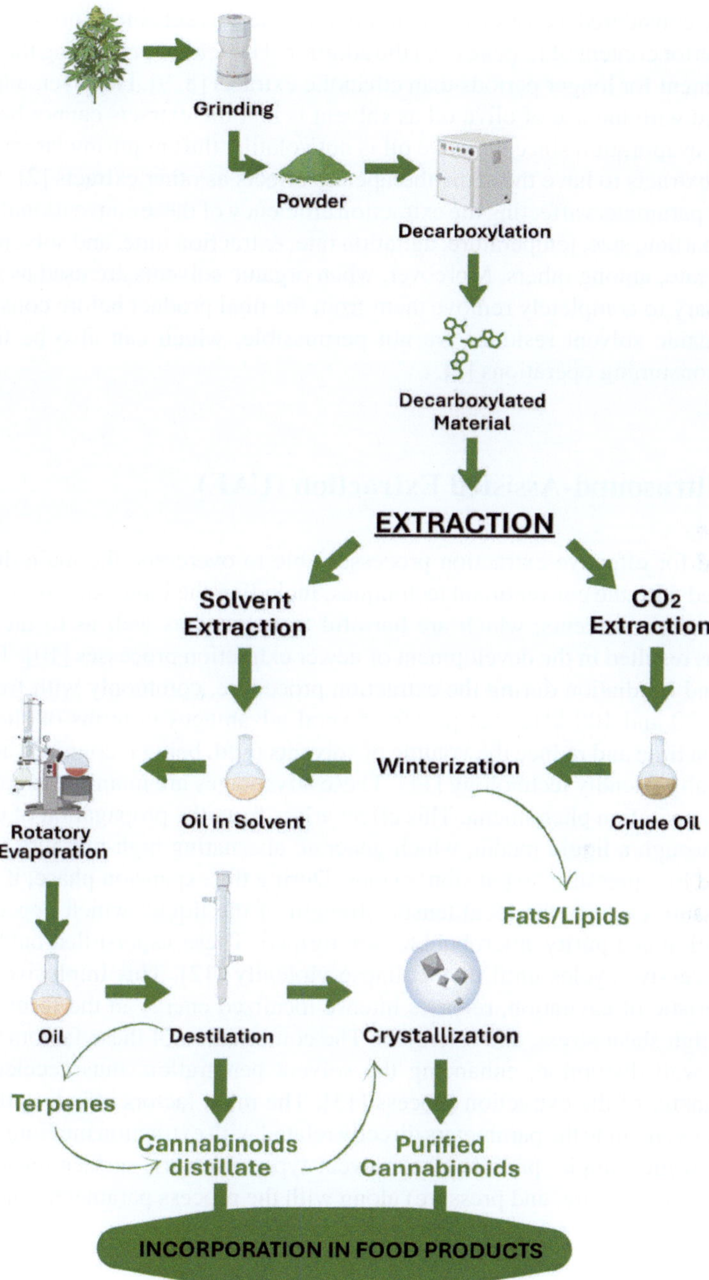

Fig. 3.1 Processing steps of the production of cannabinoids for incorporation in food products

oil can be considered a cost-effective nonflammable solvent able to produce extracts with superior content of terpenes and the additional benefit of preserving the cannabinoids content for longer periods than ethanolic extracts [8, 9]. However, a limitation associated with the use of olive oil as solvent is that the extracts cannot be concentrated by evaporation since the olive oil is not volatile, thus requiring larger volumes of these extracts to have the same therapeutic effects as other extracts [2].

Other parameters affecting the extraction efficiency of these conventional methods include particle size, temperature, agitation rate, extraction time, and solvent to plant material rate, among others. Moreover, when organic solvents are used as solvent it is necessary to completely remove them from the final product before consumption, since organic solvent residues are not permissible, which can also be time- and energy-consuming operations [2].

3.2 Ultrasound-Assisted Extraction (UAE)

The need for effective extraction processes able to overcome the main drawbacks associated with the conventional techniques, including the long extraction times and use of organic solvents, which are harmful to humans as well as to the environment, has resulted in the development of newer extraction processes [10]. The use of ultrasound irradiation during the extraction procedure, commonly with frequencies between 20 and 100 kHz, can present several advantages in terms of shorting the extraction time and reduce the volume of solvents used, being recognized as an environmentally friendly technology [11]. These advantages are mainly attributed to the acoustic cavitation phenomena. This effect arises from the propagation of ultrasonic waves through a liquid media, which generate alternating high-pressure (compression) and low-pressure (expansion) cycles. During the expansion phase, if the negative pressure exceeds the local tensile strength of the liquid, which depends on its composition and purity, microbubbles are formed. These vapor-filled bubbles grow over successive cycles until they collapse violently [12]. This implosive collapse, characteristic of cavitation, releases intense localized energy in the form of shockwaves, high shear stress, and microjets. The combination of these factors facilitates the cell wall disruption, enhancing the solvent penetration, thus accelerating the mass transfer of the extraction process [13]. The main factors affecting the extraction process include the parameters directly related with extraction medium (moisture content of the sample, particle size, solvent type, viscosity, surface tension, vapor pressure, temperature, and pressure) along with the process parameters such as time of sonication, frequency, and ultrasonic power [6, 13].

A recent review article has demonstrated that UAE is an effective method for extracting cannabinoids, offering significant improvements over conventional techniques [3]. Studies have shown that by optimizing variables, such as solvent type, extraction time, and ultrasonic power, UAE increased cannabinoids yields by 20–30%, reaching extraction efficiencies of 70–85%. Compared to conventional methods, which typically yield 50–60%, UAE not only enhances yields but also improves selectivity by reducing the co-extraction of undesired components [3]. UAE often requires specific equipment, which can increase the initial expenses, but its superior extraction performance and potential for integration with other methods make it a promising and scalable technology for recovery of cannabinoids [14]. Other studies suggest that UAE can be also useful as a conditioning step for conventional extraction methods to improve the recovery of target compounds [2].

3.3 Pressurized Liquid Extraction (PLE)

The PLE method applies high pressures to maintain the solvent at the liquid state, even above its boiling point at ambient pressure, thus increasing the extraction efficiency. Other terms often used to refer to PLE include pressurized fluid extraction (PFE), accelerated solvent extraction (ASE), pressurized solvent extraction (PSE), and enhanced solvent extraction (ESE). When water is the primary solvent, additional terms are used to refer to PLE, such as high-temperature water extraction (HTWE), superheated water extraction (SHWE), hot water extraction (HWE), and subcritical water extraction (SWE) [3].

The combination of high pressure and temperatures reduces the extraction time and the solvent requirement, making PLE a potential technique for the recovery of several bioactive compounds from plant matrices in a more environmentally friendly way [13]. The higher temperature can increase the analyte solubility and also decrease the viscosity and surface tension of the solvent, while the high pressure can facilitate the solvent penetration into the plant matrix, thus improving the extraction rate [6].

Studies demonstrated that SWE enable the recovery of a wide range of bioactive compounds from hemp, including cannabinoids, while can also be highly effective in decarboxylation of CBDA to CBD [15, 16]. In addition to water, ethanol is commonly used as solvent for PLE of cannabinoids from hemp and hemp residues, since it offers greater solubility for cannabinoids than water, resulting in improved extraction efficiency and broader recovery of bioactive compounds [3, 7]. The polarity of ethanol can be additionally tuned by adjusting its concentration with water, allowing selective extraction of target compounds. Recommended process parameters for PLE with ethanol typically include pressures between 100 and 150 bar, temperatures of 50–60 °C, extraction times from 30 to 60 min, and ethanol concentrations in the range of 70–90%, which optimizes cannabinoids yield without compromising extraction efficiency. Compared to water, the use of ethanol for PLE offers higher solubility and versatility for the cannabinoids' extraction [17].

3.4 Supercritical Fluid Extraction (SFE)

SFE is a technique that uses solvents at their supercritical state—conditions where temperature and pressure exceed the critical point of the substance. In this state, the fluid exhibits unique characteristics, combining gas-like diffusivity with liquid-like solvating power. These characteristics allow the supercritical fluid to efficiently penetrate the plant matrix and selectively extract target compounds. The setup for SFE includes a solvent tank connected to a high-pressure pump to pressurize the fluid, an extraction vessel with controlled temperature where the plant material is loaded, a separator, and a collection vessel [18, 19]. Several substances have been explored as potential solvents for SFE, but carbon dioxide (CO_2) is most widely used due to is favorable properties: CO_2 reaches the supercritical state at relatively mild conditions (31 °C and 73.8 bar), is readily available, and holds the status of "generally recognized as safe" (GRAS) [20]. These characteristics make SFE an interesting method to produce bioactive extracts with less or no environmental harmful by-products, thus receiving the status of "clean technology" [3].

A typical SFE process from plant matrices involves two main stages: (i) extraction of solutes from the raw material and (ii) separation of the extract from the solvent. In the first stage, the raw material is loaded into the extraction vessel, forming a fixed bed [2]. The system's operating conditions—namely, temperature and pressure—are carefully controlled via a pressure release valve and temperature regulation system to maintain the supercritical state of the solvent. Under these conditions, the super-critical fluid uniformly percolates through the bed, allowing solutes to diffuse from the solid matrix into the solvent. This step continues until equilibrium is reached, ensuring maximum mass transfer. In the separation stage, the mixture comprised of extracted compounds and the solvent is directed to the separation vessels. There, a controlled reduction in pressure and/or temperature decreases the solvent's dissolving capacity, leading to the precipitation of the solubilized compounds, which are then collected, while the solvent is either vented or recycled for further use [19].

One of the major advantages of SFE is the ability to fine-tune solvent's properties—such as density and selectivity—through simple adjustments in temperature and pressure, enabling tailored extraction processes with high efficiency and minimal solvent residue [21]. SFE also enables the fractionation of the compounds, which can be achieved through two main approaches: by using separation vessels in series operating at distinct pressures and temperatures to selectively precipitate compounds with different physicochemical characteristics, or by dynamically adjusting process parameters during extraction to sequentially solubilize and collect different compounds. This versatility not only enhances the yield and purity of desired bioactives—such as cannabinoids—but also helps minimizing the co-extraction of unwanted compounds, ultimately contributing to a cleaner and more efficient process [1].

Under supercritical conditions, CO_2 exhibits low polarity, which can limit its ability to solubilize more polar compounds. To overcome this limitation, small amounts of co-solvents, such as ethanol, are frequently added. These modifiers enhance the solvating power of CO_2 by increasing its polarity, thereby improving the solubility of less non-polar target compounds and expanding the range of extractable substances [20, 22]. This strategy is commonly employed to optimize extraction efficiency and selectivity.

The decarboxylation process plays an important role in the efficiency of cannabinoids extraction by SFE, as their neutral forms, such as CBD, exhibit greater solubility in supercritical CO_2 compared to their acidic precursors [1]. In the natural state of the plant, cannabinoids primarily exist in their acidic forms, such as cannabidiolic acid (CBDA), for example. Decarboxylation is a thermal-inducted chemical reaction that removes a carboxyl group from these acidic molecules, converting them into their active, neutral counterparts. The efficiency of this transformation depends on both temperature and duration [18]. An interesting approach to conduct the decarboxylation was to carry it out directly within the extraction vessel at 120 °C for 45 min prior to SFE, resulting in significantly higher concentrations of CBD and THC in the extract [23].

Temperature and pressure play a critical role in the SFE process as they directly affect the CO_2 density. The extraction efficiency of specific compounds via SFE is directly related to its solubility in the supercritical fluid, which is highly dependent on the density. For instance, CBD exhibits solubility behavior in supercritical CO_2 similar to that of CBN, with the highest solubility observed at intermediate temperatures (around 50 °C). In contrast, CBG and THC reach their maximum solubility at higher temperatures (around 70 °C). These variations in solubility among the cannabinoids are primarily attributed to differences in their molecular structures and melting points [24]. The process conditions for the recovery of cannabinoids and terpenoids using SFE typically fall within a pressure range of 100 to 380 bar and a temperature range of 25–80 °C [1]. The process parameters for SFE are generally optimized on laboratory-scale experiments and offer scalability, making it a preferable choice for industrial-scale extraction of cannabinoids [1, 3].

References

1. de Aguiar AC, Vardanega R, Viganó J, Silva EK (2023) Supercritical carbon dioxide technology for recovering valuable phytochemicals from Cannabis sativa L. and valorization of its biomass for food applications. Molecules 28. https://doi.org/10.3390/molecules28093849
2. Lazarjani MP, Young O, Kebede L, Seyfoddin A (2021) Processing and extraction methods of medicinal cannabis: a narrative review. J Cannabis Res 3. https://doi.org/10.1186/s42238-021-00087-9
3. Selvaraj S, Nawfer N, Dharmawansa KVS, Ali Redha A, Rupasinghe HPV (2025) Recent advances in cannabidiol (CBD) extraction: a review of potential eco-friendly solvents and advanced technologies. Green Anal Chem 13. https://doi.org/10.1016/j.greeac.2025.100270
4. Dey M, Bera S, Tyagi P, Pal L (2025) Mechanisms and strategic prospects of cannabinoids use: Potential applications in antimicrobial food packaging—a review. Compr Rev Food Sci Food Saf 24. https://doi.org/10.1111/1541-4337.70113
5. Luque De Castro MD, Garcia-Ayuso LE (1998) Soxhlet extraction of solid materials: an outdated technique with a promising innovative future. Anal Chim Acta 369:1–10

6. Azmir J, Zaidul ISM, Rahman MM, Sharif KM, Mohamed A, Sahena F, Jahurul MHA, Ghafoor K, Norulaini NAN, Omar AKM (2013) Techniques for extraction of bioactive compounds from plant materials: a review. J Food Eng 117:426–436. https://doi.org/10.1016/j.jfoodeng.2013.01.014

7. Fathordoobady F, Singh A, Kitts DD, Pratap Singh A (2019) Hemp (Cannabis Sativa L.) extract: anti-microbial properties, methods of extraction, and potential oral delivery. Food Rev Int 35:664–684. https://doi.org/10.1080/87559129.2019.1600539

8. Romano LL, Hazekamp A (2013) Cannabis oil: chemical evaluation of an upcoming cannabis-based medicine. www.cannabis-med.org

9. Citti C, Ciccarella G, Braghiroli D, Parenti C, Vandelli MA, Cannazza G (2016) Medicinal cannabis: Principal cannabinoids concentration and their stability evaluated by a high performance liquid chromatography coupled to diode array and quadrupole time of flight mass spectrometry method. J Pharm Biomed Anal 128:201–209. https://doi.org/10.1016/j.jpba.2016.05.033

10. Shirsath SR, Sonawane SH, Gogate PR (2012) Intensification of extraction of natural products using ultrasonic irradiations—a review of current status. Chem Eng Process 53:10–23. https://doi.org/10.1016/j.cep.2012.01.003

11. Shen L, Pang S, Zhong M, Sun Y, Qayum A, Liu Y, Rashid A, Xu B, Liang Q, Ma H, Ren X (2023) A comprehensive review of ultrasonic assisted extraction (UAE) for bioactive components: Principles, advantages, equipment, and combined technologies. Ultrason Sonochem 101. https://doi.org/10.1016/j.ultsonch.2023.106646

12. Luque-García JL, Luque De Castro MD (2003) Ultrasound: a powerful tool for leaching, TrAC. Trends Anal Chem 22:41–47. https://doi.org/10.1016/S0165-9936(03)00102-X

13. Vardanega R, Santos DT, De Almeida MA (2014) Intensification of bioactive compounds extraction from medicinal plants using ultrasonic irradiation. Pharmacogn Rev 8:88–95. https://doi.org/10.4103/0973-7847.134231

14. Dzah CS, Duan Y, Zhang H, Wen C, Zhang J, Chen G, Ma H (2020) The effects of ultrasound assisted extraction on yield, antioxidant, anticancer and antimicrobial activity of polyphenol extracts: a review. Food Biosci 35 (2020). https://doi.org/10.1016/j.fbio.2020.100547

15. Olejar KJ, Hatfield J, Arellano CJ, Gurau AT, Seifried D, Vanden Heuvel B, Kinney CA (2021) Thermo-chemical conversion of cannabis biomass and extraction by pressurized liquid extraction for the isolation of cannabidiol. Ind Crops Prod 170. https://doi.org/10.1016/j.indcrop.2021.113771

16. Drinić Z, Vladic J, Koren A, Zeremski T, Stojanov N, Tomić M, Vidović S (2021) Application of conventional and high-pressure extraction techniques for the isolation of bioactive compounds from the aerial part of hemp (Cannabis sativa L.) assortment Helena. Ind Crops Prod 171. https://doi.org/10.1016/j.indcrop.2021.113908

17. Serna-loaiza S, Adamcyk J, Beisl S, Kornpointner C, Halbwirth H, Friedl A (2020) Pressurized liquid extraction of cannabinoids from hemp processing residues: evaluation of the influencing variables. Processes 8:1–16. https://doi.org/10.3390/pr8111334

18. Baldino L, Scognamiglio M, Reverchon E (2020) Supercritical fluid technologies applied to the extraction of compounds of industrial interest from Cannabis sativa L. and to their pharmaceutical formulations: a review. J Supercrit Fluids 165. https://doi.org/10.1016/j.supflu.2020.104960

19. Vardanega R, Náthia-Neves G, Veggi PC, Meireles MAA (2019) Supercritical fluid processing and extraction of food. In: Green food processing techniques. Elsevier, pp 57–86. https://doi.org/10.1016/B978-0-12-815353-6.00003-3

20. Rovetto LJ, Aieta NV (2017) Supercritical carbon dioxide extraction of cannabinoids from Cannabis sativa L. J Supercrit Fluids 129:16–27. https://doi.org/10.1016/j.supflu.2017.03.014

21. Vardanega R, Osorio-Tobón JF, Duba K (2022) Contributions of supercritical fluid extraction to sustainable development goal 9 in South America: industry, innovation, and infrastructure. J Supercrit Fluids 188. https://doi.org/10.1016/j.supflu.2022.105681

22. Ribeiro Grijó D, Vieitez Osorio IA, Cardozo-Filho L (2019) Supercritical extraction strategies using CO_2 and ethanol to obtain cannabinoid compounds from cannabis hybrid flowers. J CO_2 Utilization 30:241–248. https://doi.org/10.1016/J.JCOU.2018.12.014

23. Fernández S, Carreras T, Castro R, Perelmuter K, Giorgi V, Vila A, Rosales A, Pazos M, Moyna G, Carrera I, Bollati-Fogolín M, García-Carnelli C, Carrera I, Vieitez I (2022) A comparative study of supercritical fluid and ethanol extracts of cannabis inflorescences: chemical profile and biological activity. J Supercrit Fluids 179:105385. https://doi.org/10.1016/J.SUPFLU.2021.105385

24. Perrotin-Brunel H, Kroon MC, Van Roosmalen MJE, Van Spronsen J, Peters CJ, Witkamp GJ (2010) Solubility of non-psychoactive cannabinoids in supercritical carbon dioxide and comparison with psychoactive cannabinoids. J Supercrit Fluids 55:603–608. https://doi.org/10.1016/J.SUPFLU.2010.09.011

Chapter 4
Analytical Techniques for Identification and Quantification of Cannabinoids

The rapid expansion of the cannabis industry, driven by both medical and recreational markets, has introduced a wide variety of cannabis and cannabis-based products. This increasing diversity of products requires a rigorous market surveillance to ensure the safety of patients and consumers [1]. Among the numerous bioactive compounds present in these products, cannabinoids are the main compounds of interest due to their pharmacological relevance. Historically, the analytical methods were mainly focused on quantification of THC. However, over the years, other cannabinoids have demonstrated pharmacological properties, and the quantification of these compounds became fundamental to understanding the biological properties of cannabis [2]. As a result, there is a growing demand for robust qualitative and quantitative analytical methods to characterize cannabinoids content in different areas [3].

These analytical efforts are essential not only for quality control purposes but also for distinguishing between industrial fiber hemp and THC-cannabis intended for recreational use [4]. Obviously, the quality control analyses of cannabis and cannabis-derived products are broader than the quantification of cannabinoids potency since it also involves microbiological, pesticide, heavy metal and aflatoxin contamination, as well as moisture content determination. In addition, tests to comply with good agricultural and manufacturing practices are required for the products intended for medical purposes [3].

Although numerous analytical methods have been developed for cannabinoid analysis, only a limited number have undergone validation in accordance with international recognized standards while researchers are still looking for improved methodological and technological solutions [3, 5]. Several challenges make the implementation of robust and standardized analytical procedures for cannabinoids quantification in cannabis-derived products difficult. These include the determination of appropriate level of specificity among different botanical matrices, assess method's accuracy (recovery), and ensure inter-laboratory precision (reproducibility), all within the context of strict and dynamic regulatory environments [5]. This section

S. E. Azam et al., *Cannabinoids in Food Science*,
Chemistry of Foods, https://doi.org/10.1007/978-3-032-11546-1_4

mainly focuses on the analytical techniques employed for the quantification of cannabinoids in plant materials and cannabis-derived products.

4.1 Sample Preparation

The sample preparation is a critical step in ensuring the reliable determination of cannabinoids across different samples. This is considered the most labor-intensive and time-consuming analytical step and accounts for 60–80% of the total analysis time, often representing the bottleneck of the entire analytical procedure [6, 7]. The complexity and composition of the sample matrix significantly affects key analytical parameters such as recovery rates, accuracy, and reproducibility. As a result, preparation protocols optimized to one type of sample are often not directly transferable to others, highlighting the need for matrix-specific method's development and validation [8].

4.1.1 Plant Material

Cannabis plant is complex and heterogeneous matrix, as its different parts contain different cannabinoid profiles [9]. The plant material preparation for cannabinoid quantification typically focuses on the inflorescences, as these are the primary site of cannabinoid accumulation in the plant. Sample preparation of plant materials commonly involves grinding the material to reduce particle size, thereby increasing surface area and improving extraction efficiency, followed by a solid–liquid extraction process.

Although most of the active compounds are located on the superficial granular trichomes of cannabis plant, a significant portion is also distributed in non-granular tissues. It highlights the importance of particle reduction during sample preparation, as it enhances the efficiency of subsequent extraction and recovery of cannabinoids [10]. The particle reduction can be achieved by manual procedures, such as pulverization using mortar and pestle, or by mechanical grinding using laboratory mills (i.e., knife or ball mill) [9, 11]. However, the mechanical grinding can affect the recovery of total cannabinoids due to adhesion of cannabinoid-rich resin to the blades and plastic surfaces of the grinder during high-speed pulverization [12].

Some methods also recommend drying the material before grinding to residual moisture values in the range of 8–13% [11, 13]; however, it requires an additional moisture determination to accurately assess the initial moisture content of the plant material [2]. Several approaches have been used for drying the plant material, but mild conditions of temperature around 35–40 °C for 24–48 h are generally preferred to avoid compound degradation and terpenes volatilization [9]. Sieving the material after griding is also recommended to ensure homogeneity of the sample [11].

The dried and comminuted samples are then subjected to solid–liquid extraction to recover the cannabinoids prior to analysis. To this end, acid and neutral cannabinoids can be extracted using organic solvents or mixtures of them. The most common solvent is ethanol, which has a strong affinity for the cannabinoid's molecular structure, leading to high extraction efficiency [2, 4, 9, 14]. In addition, ethanol satisfies green chemistry principals, since it is recognized as a GRAS solvent. However, other solvents such as methanol, ethyl acetate, and hexane are also widely applied alone or in combination with other solvents [4]. For instance, UNODC [11] and the American Herbal Pharmacopoeia [13] recommend the extraction with methanol:chloroform (9:1, v/v), while the German Pharmacopoeia [15] propose ethanol 96% (v/v) as solvent. Although cannabinoids present a high affinity with ethanol, this solvent co-extract significant amount of other compounds enhancing matrix interference [10]. To minimize this effect, hexane can be alternatively used [16, 17].

As previously discussed in Chap. 3, cannabinoids can be extracted from the cannabis plant using several extraction techniques. However, for sample preparation purposes, maceration and UAE are among the most widely employed methods. When the analytical focus is to quantify the acidic cannabinoids, the extraction should be performed at room temperature to prevent thermal decarboxylation, thereby preserving the native profile of cannabinoids in the plant material [4]. In contrast, the preparation of extracts for medicinal purposes requires the presence of pharmacologically active neutral cannabinoids, which can be achieved by either extraction at higher temperatures or preliminary decarboxylation step [4]. To minimize cannabinoid loss during decarboxylation, the reaction should be carried out in closed rectors, under elevated temperature and short-duration conditions [18].

4.1.2 *Cannabis-Derived Products*

The global market for cannabis-derived products, especially edibles, has grown rapidly, with an increasing variety of commercially available items. These products cover a wide range of food matrices, including beverages such as juices, tea, and coffee-based drinks, as well as solid and semi-solid products, such as ice cream, honey, cookies, candies, gummies, and chocolates [19]. This diversity reflects a dynamic and evolving market, driven by shifting consumer preferences and ongoing changes in cannabis legislation [20].

Despite their rising popularity, cannabis-derived edibles present significant regulatory challenges, particularly in ensuring accurate quantification of cannabinoids and identification of potential contaminants [20, 21]. These concerns are especially critical in the context of public health, given the risk posed to vulnerable populations such as children and individuals with specific medical conditions [20]. A primary obstacle to reliable analysis of cannabinoids lies in the complexity of extracting these compounds from diverse food matrices. The presence of fats, sugars, proteins, carbohydrates, and pigments in these products often interferes with standard analytical procedures, making it necessary to employ tailored extraction protocols to ensure efficient cannabinoids recovery. Consequently, the choice of extraction method is strongly dependent on the specific composition of the edible matrix.

Ciolino et al. [22] conducted a comprehensive qualitative GC–MS analysis of over 60 cannabis-derived products, including oral supplements, foods, candies, beverages, vapes/e-liquids, and topical formulations. To minimize matrix interferences, the authors evaluated various ethanol- and acetonitrile-based extraction solvents. In most cases, 95% ethanol was suitable as extracting solvent; however, significant interferences were observed in samples with high content of glycerin, sugar/carbohydrates, and/or lactose.

While sugars and glycerin generally do not interfere with HPLC–DAD analysis, these compounds can significantly impact GC–MS results. These compounds compete with cannabinoids for derivatization agents, leading to incomplete derivatization of CBD, and promoting undesired side conversions to Δ^8-THC and Δ-THC. To reduce the co-extraction of glycerin, the authors used acetonitrile as extracting solvent, given the poor solubility of glycerin in acetonitrile (less than 5%) [22]. Acetonitrile-based extracting solvents were also effective in minimizing the co-extraction of sugars in products with high content of carbohydrates [22].

For lactose-containing products, such as Greek yogurt, cream cheese, butter, and milk-based coffee beverages, a pre-treatment with lactase was incorporated prior to extraction using acetonitrile-based solvents. This enzymatic step helped hydrolyzing lactose, thereby minimizing its analytical interference. In the case of semi-solid fatty/oily matrices (e.g., butter, margarine, chocolate bars, and non-polar topic ointments or balms), a warming step was applied after solvent addition to facilitate melting and homogenization of the samples. By adopting these tailored sample preparation methods to the specific characteristics of each matrix, cannabinoids recoveries ranged from 58 to 111% [22].

A recent review compiled the information of HPLC–UV approaches for cannabinoid analysis from 2022 to 2024 [19]. The authors observed that most of the recent articles describing analysis of cannabinoids in food products were focused on gummy matrices. In these cases, the methods involved UAE, or shaking extraction using water or aqueous mixtures with methanol, acetonitrile, or MTBE.

In addition to tailoring extraction protocols, more advanced techniques focusing on clean-up of the samples, including solid-phase extraction (SPE), QuEChERS, dissolution and dispersion, and liquid-phase extraction, have been successfully employed for isolation and analysis of cannabinoids from edible matrices [20]. Among them, QuEChERS, which stands for "Quick, Easy, Cheap, Effective, Rugged, and Safe," has gained increased popularity in analyzing several compounds, including cannabinoids in complex food matrices [23–25]. QuEChERS is a simple extraction and purification method based on the principal of partitioning analytes between a water-miscible organic solvent and an aqueous phase, typically involving acetonitrile as the solvent and the addition of salts like magnesium sulfate ($MgSO_4$) and sodium chloride (NaCl) to induce phase separation [20, 26].

Lindsay et al. [24] applied a modified QuEChERS extraction method in 45 cannabis-derived edibles collected from the Jamaican market, prior to GC–MS analysis. The samples included baked goods, candies, chocolates, and ice cream. After grinding the samples to a fine powder, they were weighted and mixed with distilled water at a solid-to-liquid ratio of 1:10 (w/v) followed by vortexing for 30 s. Then 10 mL of acetonitrile with 1% acetic acid was added, the mixture was vortexed for 1 min, and shaken at 150 rpm for 1 h. The extraction salts (MgSO4 + NaCl, 4:1, w/w) were added and the mixture was vortexed for 1 min, centrifuged at 3000 rpm for 5 min, and the supernatant transferred to a tube. The solvent was then evaporated to dryness at 40 °C, under gentle stream of nitrogen and the dried extract was reconstituted in 1 mL of hexane:ethyl acetate (1:1, v/v). The recovery of cannabinoids for all samples ranged from 96 to 114%, demonstrating the efficiency of the QuEChERS. Interestingly, the results demonstrated that 30% of the samples had THC levels greater than the recommended 10 mg THC per serving, raising the public health concerns for all consumers including inexperienced users who may be at a great risk of overdosing.

Although QuEChERS is celebrated for its simplicity, speed, and cost-effectiveness, this method may not provide the same level of selectivity and purity as more sophisticated methods such as SPE [20]. Thus, approaches also involve an additional clean-up step using SPE with several combinations of porous sorbents and salts to remove interferents [26].

4.1.3   Biological Matrices

As new cannabis-derived products and technologies continue to be developed and tested, the variety of sample of matrices has expanded, driven by the growing number of both in vitro and in vivo tests. Cannabinoids have been extracted from a wide range of biological matrices, including blood and plasma, urine, oral fluid, hair, and different tissues, collected from both humans and diverse animal models [19]. he quantification of cannabinoids in biological samples is primarily conducted to support clinical studies and, additionally, for forensic purposes to detect the illicit use of cannabis-derived products [4]. In this sense, the development of analytical methods for quantifying cannabinoids and their metabolites in biological samples is imperative for research to advance toward the therapeutic effects of cannabinoids, also supporting the development of safe food products containing cannabinoids. Furthermore, in the forensic field, robust, sensitive, and reliable techniques are crucial for accurately analyzing biological samples from individuals, thereby avoiding the risk of false positives [4, 6]. Due to the complexity of biological tissues and fluids, robust sample preparation techniques are required to eliminate the interferences while selectively extracting and pre-concentrating the target analytes, which are often present at trace levels in biological matrices [6, 7].

As proteins are the primary interferent present in biological samples, protein precipitation is usually the first step evaluated when developing analytical methods to quantify cannabinoids and their metabolites in biological matrices [4, 6, 7]. Several precipitants can be used for this purpose, including pure or mixed organic solvents (e.g., methanol, acetone, and acetonitrile), acidic agents (e.g., H_2SO_4, CF_3COOH, $ZnSO_4$, $(NH_4)_2SO_4$, NH_4NO_3, NH_4Cl, CCl_3COOH, and $HClO_4$), and neutral salts (e.g., $MgSO_4$, Na_2SO_4, NaCl, $MgCl_2$, CH_3COONH_4, and $HCOONH_4$) [27]. In some cases, enzymatic treatments have been also explored for effective deproteinization [4]. Following deproteinization, samples are typically subjected to additional preparation steps such as liquid–liquid extraction (LLE), solid–liquid extraction (SLE), SPE, and solid-phase microextraction (SPME) [4, 6, 7].

LLE is based on the principle of partition equilibrium, where the analyte distributes between two immiscible solvents—typically an aqueous solution and an organic solvent. The preferred solvents for LLE of cannabinoids include ethyl acetate, dichloromethane, chloroform, toluene, hexane, cyclohexane, and their mixtures [6]. Although this is a simple preparation technique that provides good repeatability and high recovery for most cannabinoids, it presents some drawbacks related with the difficult automation, large volumes of solvents required in addition to be time-consuming, and environmentally unfriendly, in contrast with the principles of green chemistry [28–30]. For the preparation of solid biological matrices, such as hair, nails, and organ tissues, SLE procedures are employed as pre-treatment in conjunction with other techniques, usually SPE to increase selectivity and extraction efficiency. For hair and nails, a washing step with organic solvents is recommended to remove possible contaminants [6].

SPE has been preferred for the preparation of biological matrices due to its high reproducibility, easy automation, shorter extraction time, and less solvent consumption compared to LLE [4, 6]. In SPE, the target analytes are absorbed onto a stationary solid-phase material, reducing the presence of interferents in the sample, allowing a pre-concentration of the analytes before analysis and improving detection sensitivity [4, 31]. Depending on the interactions that occur between the analyte and the sorbent, commercial SPE adsorbents are grouped into reverse phase (hydrophobic interaction), ion exchange (electrostatic interactions), and mixed mode (hydrophobic and electrostatic interactions) [32]. However, the main limitation of SPE methods includes the high cost of the extraction cartridges that are of single use only [6]. The solvents typically used to elute cannabinoids from SPE do not differ from those used in LLE, such as methanol, ethanol, hexane, and ethyl acetate, and in some cases added with acetic acid, depending on the composition of the SPE stationary phase [4]. More recently, SPME devices, which are miniaturized alternatives to conventional SPE, have been proposed to achieve efficient extraction of target analytes while reducing solvent consumption and preparation time [33, 34].

A recent review on advances in chromatographic analysis of endocannabinoids and phytocannabinoids in biological samples concluded that, although SPME is the fastest sample preparation method, it is also the most technically complex and costly. In contrast, LLE is the most widely used due to its simplicity and lower cost compared to SPE. However, SPE offers greater efficiency in removing interferences and minimizing matrix effect, which may justify its preference in more demanding analytical contexts [7]. The QuEChERS approach has also emerged as a more environmentally friendly and cost-effective alternative to conventional sample preparation techniques of biological matrices [6].

4.2 Analytical Techniques

The analytical platforms for profiling cannabinoids in cannabis and its derived products have advanced significatively over the recent decades, driven by the rapid scientific and technological progress of cannabis-based formulations. Developing reliable analytical methods for cannabinoids determination across various matrices presents considerable challenges, particularly due to the chemical nature of these compounds [9]. Cannabinoids are biosynthesized in the plants primarily in their acidic forms, which can be converted into their neutral counterparts through exposure to heat or light. Therefore, accurate quantification of both acidic and neutral cannabinoid forms requires analytical approaches that carefully control temperature and light exposure, as well as detection techniques capable of distinguish between them [35].

Chromatographic techniques—particularly liquid chromatography (LC) or gas chromatography (GC) coupled with mass spectrometry (MS)—are considered the gold standard for cannabinoid analysis and are widely employed in both research and routine analyses contexts [9, 35, 36]. In addition to MS, other detectors such as diode array detectors (DAD) in LC and flame ionization detectors (FID) in CG are also commonly used, depending on the required sensitivity and specificity [35]. For preliminary or qualitative screening, thin-layer chromatography (TLC) remains a useful tool due to its simplicity and cost-effectiveness [4, 35].

4.2.1 Gas Chromatography

GC is one of the most widely used techniques for the analysis of cannabinoids in both plant-derived materials and biological matrices [4, 35]. Actually, GC is the officially recognized method by authorities for cannabinoids determination in forensic applications, particularly for the detection of illicit cannabis use [3]. Typically, GC offers several analytical advantages, including lower operational costs, shaper chromatographic peaks, and greater peak capacity compared to LC [6].

However, a notable limitation of GC in cannabinoid analysis lies in the high temperatures required for sample volatilization, which induce the thermal decarboxylation of acidic cannabinoids into their correspondent neutral forms. This transformation compromises the accurate quantification of acidic cannabinoids. Therefore, a derivatization step of the cannabinoid acids is typically required prior to GC analysis. This step stabilizes the acidic forms of cannabinoids, allowing their detection and quantification without thermal degradation, and thereby enabling differentiation between neutral and acidic cannabinoids [4, 35, 36]. However, complete derivatization of cannabinoid acids is difficult to obtain, leading to inaccurate results as thermal degradation can occur in the GC analysis [4, 37, 38]. In addition, the necessity of derivatization steps increases the analyses costs and time [9].

Siltation and methylation are the most common derivatization reactions employed for cannabinoids [6]. It involves the substitution of a hydrogen atom that is bound to a hetero-atom (such as –OH, –COOH, –NH2, =NH, and SH) by a silyl group, i.e., trimethylsilyl (TMS) or tert-butyldimethilsilyl (TBDMS) in the silytation or a methyl group in the case of methylation [9]. Commonly used derivatization reagents include acid anhydride, N-tert-butyldimethylsilyl-N-methyltrifluoroacetamide (MTBSTFA), n-propyl iodide (C_3C_7I), N,O-bis(trimethylsilyl)trifluoroacetamide (BSTFA), N-methyl-N-(trimethylsilyl)trifluoroacetamide (MSTFA), hexafluoroisopropanol (HFIP), and methyl iodide (CH_3I) [39]. However, acid derivatization techniques selected with caution, as they may promote undesirable chemical transformations. For instance, it has been reported that approximately 84% of CBD can convert into Δ^9-THC and Δ^8-THC under acidic conditions. Additionally, 7-COOH-CBD may can be converted to 11-COOH-THC [40]. Such conversions not only compromise analytical accuracy but may also lead to false interpretations of cannabinoid content.

A typical derivatization procedure consists of incubating the dried sample—obtained after clean-up steps such as LLE and SPE—with excess of derivatization reagent at temperatures ranging from 60 to 100 °C for 20 to 40 min [6]. For example, a dried cannabinoid eluate can be derivatized using BSTFA/trimethylchlorosilane (TMCS) mixture (99:1, v/v) at 100 °C for 20 min. An innovative alternative reported by Rosendo et al. [41] demonstrated that microwave-assisted derivatization with MSTFA/TMCS at 800 W can dramatically reduce the reaction time to 2 min.

In GC analysis of cannabinoids, column selection plays a critical role in achieving optimal separation, resolution, and sensitivity. Most analytical methods target the major phytocannabinoids, including Δ^8-THC, Δ^9-THC, Δ^9-THCA CBC, CBD, CBDA, CBDV, CBG, CBGA, CBN, and THCV [9]. These compounds contain aromatic rings, alkyl chains, and hydroxyl groups, which interact differentially with stationary phases depending on polarity and phenyl group content. Consequently, the proportion of phenyl groups in mixed dimethylpolysiloxane-silphenylene or dimethylpolysiloxane-diphenyl stationary phases can significantly influence separation efficiency and selectivity [9]. Among non-polar columns, 5%-diphenyl-dimethylpolysiloxane capillaries such as HP-5 (commonly used with FID) and HP-5MS (for MS) are widely employed for general cannabinoid profiling [6, 9, 42, 43]. For more refined separations—particularly of structurally similar cannabinoids like CBC and CBD—100% dimethylpolysiloxane columns (e.g., HP-1, OV-1, and DB-1) have shown improved performance, either isolating these two compounds from each other or from other phytocannabinoids [44–47].

Columns of intermediate polarity are also utilized. For instance, Rxi-35Sil MS and DB-35MS, which contain 35% (phenyl)methylpolysiloxane, offer a broader retention window than columns with lower phenyl content (e.g., HP-5MS or DB-5MS) [9, 22]. However, a limitation of these columns is the potential for peak splitting or coelution, particularly involving CBC and THCV, which may compromise quantification accuracy [9]. In contrast, columns with low-to-mid polarity such as DB-1701 (14% cyanopropylphenyl–methylpolysiloxane) are generally considered suboptimal, as they may yield poor responses, distorted peak shapes, and significant tailing [9]. Interestingly, highly non-polar columns like the DB-1HT (100% dimethylpolysiloxane) have demonstrated successful separation not only for major cannabinoids, such as Δ^9-THC, CBD, CBC, and CBN, but also for minor cannabinoids, terpenes, sesquiterpenes, sterols, diglycerides, and triglycerides [48].

Regarding detectors, FID and MS are the most commonly used in cannabinoid analysis [4, 42]. FID offers highly accurate quantitative results without the need for costly deuterated internal standards, making it advantageous for routine quantification. However, its sensitivity is notably lower than that of GC–MS—typically just below 1 μg/mL compared to detection limits in the order of, or even below, 1 ng/mL for GC–MS [4]. Moreover, MS provides higher specificity than FID, particularly when chemically similar compounds co-elute, and is indispensable when structural elucidation or enhanced sensitivity is required—for example, in bioanalytical applications or for profiling minor cannabinoids [3]. For quantitative analysis using MS, the use of internal standards—preferably deuterated analogs—is strongly recommended to reduce signal variability and correct for potential matrix effects on analyte ionization (such as ion suppression or enhancement). However, these deuterated standards are commercially available only for the major cannabinoids, limiting their use for comprehensive cannabinoid profiling [3, 4].

Some studies have indicated that one-dimensional GC does not provide enough resolution in the separation of cannabinoids due to their similar chemical structures [49, 50]. Hence, two-dimensional (GC × GC) techniques have been employed to provide a more efficient evaluation of chemical profile of such complex matrices [51, 52]. GC × GC is an advanced separation technique that uses two columns with different stationary phase properties connected in sequence. A modulator traps and injects small amounts of sample from the first column into the second, hence improving peak capacity, resolution, and the ability to analyze complex mixtures by spreading overlapping peaks into two dimensions for clearer identification and quantification. Omar et al. [53] used a GC × GC/qMS method to resolve co-eluted compounds in cannabis extracts. In this case, the first dimension consisted of HP-5MS capillary column and the second dimension consisted of DB-17MS, i.e., a non-polar stationary phase in the first dimension, followed by a medium polarity in the second dimension. A similar method GC × GC/qMS using the same columns set was also applied for the analysis of cannabinoids in post-mortem blood samples [51]. Although the two-dimensional GC offers some advantages over the classical one-dimensional GC—such as signal enhancement, higher separation capacity, and structure-retention dimensionality—it requires specific and specialized approaches to analyze experimental data and to extract the chemical information [53].

4.2.2 Liquid Chromatography

Recently, LC methods have gained increasing prominence for cannabinoids analysis across diverse matrices over GC methods [3, 6]. Although GC–MS/MS has been the main choice over the last decades, preference for LC–MS/MS has been evident in the literature over the last 2 years [6]. In contrast to GC, LC eliminates the need for derivatization, thereby reducing analysis time and avoiding the additional costs and complexities associated with this step. In addition, LC allows the direct analysis of acidic cannabinoids, with a reduced risk of thermal degradation (e.g., oxidation, isomerization) [35, 38].

Among the LC techniques, ultra-performance liquid chromatography (HPLC) and ultra-performance liquid chromatography (UPLC) are the most prominent. Compared to HPLC, UPLC operates with columns packed with smaller particles sizes (≤ 2 µm), thus requiring significantly higher pressures. While the use of conventional HPLC methods still dominates the analysis of cannabinoids, a steady increase in the methods using UPLC in profiling cannabinoids across diverse matrices has been observed [52]. This increase in the use of UPLC methods is mainly associated with its advantages over HPLC, such as increased efficiency, for faster run times, sharper peaks, and enhanced resolution. Additionally, UPLC generally provides greater sensitivity and reduces solvent consumption, making it cost-effective and environmentally friendly [7].

A variety of columns are available for HPLC and UPLC analysis of cannabinoids, but most methods employ similar stationary phases. Reverse-phase C_{18} packed columns are the most common choice, followed by C_8 or phenyl columns [52]. Popular examples include Ascentis Express C18, Waters Cortecs Shield RP-18, InfinityLab Poroshell 120 EC-18, NexLeaf CBX for Potency C18, Phenomenex Kinetex C18, and Phenomenex Luna C18 [19]. An important advancement in this field is the adoption of superficially porous particles—also known as core–shell or fused-core particles—which substantially columns performance [54], by combining the benefits of both fully porous and nonporous particles [3]. Currently, this particle technology dominates the columns used for cannabinoids separation.

While C_{18} phases are the most prevalent, subtle differences in stationary phase chemistry can influence polarity and, consequently, cannabinoids separation. For example, Poroshell 120 EC-C18, Kinetex C18, and Ascentis Express C18 use dimethyl-octadecyl phases; Raptor ARC-18 features a diisobutyl-octadecyl phase, and Cortecs Shield RP-18 and Ascentis Express RP-Amide incorporate embedded polar functional groups. These variations can fine-tune selectivity and may offer performance advantages depending on the specific analytical goals [19]. The use of guard columns with the same stationary phase is recommended to prevent the column from contamination and premature degradation, especially when analyzing complex matrices [5, 19, 54].

As mobile phases, most studies report the use of acetonitrile (ACN), methanol (MeOH), and water containing small percentages, usually 0.1%, of formic or acetic acid [52]. Gradient elution is generally preferred, as it enables shorter run times, high flow rates, and improved peak resolution. However, successful chromatographic separation using isocratic elution has also been reported [19, 55]. Among organic solvents, ACN is often favored over MeOH due to its lower viscosity, which reduces system backpressure and allows higher flow rates, as well as its lower UV absorbance at wavelengths commonly used for cannabinoids analysis, leading to greater method sensitivity and baseline stability. However, each solvent offers unique properties that can be advantageous depending on the separation requirements. For instance, MeOH exhibits mild acidic properties, whereas ACN exhibits dipolar properties, and some studies have reported enhanced cannabinoids separation when CAN:MeOH mixtures are employed [19]. Water acidified with 0.1% formic acid is the aqueous phase most widely used. The addition of pH modifiers not only enhances selectivity by increasing ionic strength but also improves peak shape with minimum tailing [56, 57]. When MS detection is used, the addition of formic acid can enhance ionization of analytes [7].

A variety of detection strategies can be coupled to HPLC for the analysis of different compounds. For cannabinoids, ultraviolet (UV)-based detection in the range of 190–400 nm is most commonly employed, owing to the presence of chromophores in their chemical structures and the linearity of UV responses [3]. This approach is straightforward and cost-effective [43], but requires adequate chromatographic separation prior to detection, as UV detectors cannot distinguish between co-eluting compounds with similar absorption profiles [3, 5]. In contrast, diode array detector (DAD), which covers the visible and UV spectra, can enhance specificity because acidic and neutral cannabinoids exhibit distinct absorption patterns [42, 43]. Acidic cannabinoids display a major absorption peak near 220 nm, along with two minor peaks around 260–270 and 300 nm, whereas neutral cannabinoids typically present a single peak around 210 nm [9, 42, 58]. These spectral differences can be leveraged to improve method sensitivity, although single-wavelength detection—most often set between 210 and 240 nm, with 220 nm as the preferred choice—remains common [9].

When improved selectivity is required, MS detectors can be employed, as they can discriminate cannabinoids based on the mass-to-charge ratio (m/z) value of their molecular ions [4]. However, because m/z values are not always unique, distinguishing isomeric compounds can be challenging [42]. High-resolution mass analyzers overcome this limitation by providing accurate mass measurements of precursor and fragment ions, enabling confirmation of elemental composition [7]. In this respect, time-of-flight analyzers coupled with quadrupole mass filter (q-ToF) and ion trap systems, such as Orbitrap, can reliably differentiate compounds with identical nominal masses but different elemental compositions [3, 4, 55]. For quantitative applications, triple-quadrupole (QqQ) instruments are widely used due to their excellent sensitivity and selectivity [7]. However, as QqQ systems provide only nominal mass measurements without structural identification, the use of deuterated internal standards is mandatory to achieve accurate quantification and minimize matrix effects, which can significantly influence MS response [3, 4].

The selection of the appropriate detector largely depends on the analytical target. For pharmaceutical samples, where qualitative composition is expected to remain consistent, UV/DAD detectors may be sufficient. In contract, forensic samples often exhibit significant qualitative variability over time and across sources. Such variations can produce unexpected peaks, potentially compromising method selectivity and accuracy during routine analysis [3].

4.2.3 Supercritical Fluid Chromatography

Supercritical fluid chromatography (SFC), first developed in the 1960s, has not yet reached the widespread adoption of LC and GG for the analysis, separation, and quantification purposes [9, 54]. Leveraging the intermediate properties of supercritical fluids—combining the high diffusivity of gases with the solvating power of liquids—SFC offers excellent kinetic performance and good analytes solubility for analyzing a variety of substances [3, 54]. Historically, SFC found its main chiral separations due to its compatibility with chiral stationary phases originally designed for normal-phase LC, combined with the low viscosity and high diffusivity of supercritical CO_2, which enable high flow rates and resolution [59]. This strong performance, together with extensive adoption by the pharmaceutical industry for high-throughput chiral screening in the 1980s–1990s, reinforced its reputation as a niche chiral technique—a perception only recently challenged by advances in ultra-high performance SFC (UHPSFC) for achiral applications [59].

The renewed interest in SFC has been driven by innovations in mobile phases composition, the development of novel stationary phases, and the availability of modern SFC systems modeled after UHPLC instruments, which provide markedly improved reliability and performance [60]. The use of supercritical CO_2 as the primary mobile phase enables the replication of normal-phase conditions without employing toxic organic solvents, aligning well with green chemistry principals [3, 61]. The introduction of sub-2μm particle columns—similar to UHPLC—has further enabled highly efficient chromatographic separations to be achieved in very short analysis times [40].

When coupled to UV or MS detection, UHPSFC has proven to be a powerful tool for cannabinoids analysis [9]. Studies demonstrate that SFC can deliver rapid (8–10 min run time), highly specific, and high-resolution cannabinoids profiling [61, 62]. Unlike GC, it does not require prior derivatization and/or decarboxylation, and it can simultaneously separate neutral from the acidic cannabinoids. Furthermore, UHPSFC is considered a highly orthogonal technique, as it can alter the elution order and relative retention time of the compounds compared to UHPLC, thereby enhancing the discrimination of cannabinoids in complex matrices, particularly when combined with MS [9].

Most of SFC reports regarding chiral separation of cannabinoids focus on polysaccharide-based chiral stationary phases (both amylose and cellulose), which have been the most widely used for this goal [63–65]. Regarding the detectors coupled to SFC, the same considerations made for LCA are valid, which successfully use UV and MS detectors [3]. For MS hyphenation to SFC, the presence of splitter and makeup pumps in the interface is necessary to achieve a sufficient degree of flexibility and high detection sensitivity in electrospray ionization [54, 66].

Based on the above it is evident that the optimal separation and identification of cannabinoids from plant matrices or cannabis-derived products requires selecting from chromatographic techniques with distinct strengths and limitations. Therefore, the choice of analytical method should be guided by the specific objectives of the analysis and the technical capabilities of the laboratory [35].

4.2.4 Other Techniques

Alternative analytical techniques to conventional chromatographic methods for cannabinoid determination include nuclear magnetic resonance (NMR) spectroscopy, vibrational spectroscopy based on infrared (IR), and Raman spectrophotometers [9, 67] These techniques have gained prominence as process analytical tools in the pharmaceutical industry for monitoring various quality attributes, owing to their capacity for high-throughput screening and analysis of large sample volumes within short periods of time. They offer rapid, versatile, and non-invasive approaches for both qualitative and quantitative profiling, as wells as for assessing growth stage of cannabis plant and characterizing extracts [9].

NMR spectroscopy has been used as a semi-quantitative technique either as a standalone method or as an orthogonal approach alongside LC or GC for both qualitative and quantitative purposes [68–70] Quantitative NMR is recognized for its high accuracy, reproducibility, and relatively short analysis time. A notable advantage over LC and GC is its lack of sensitivity toward common plant impurities, such as chlorophyll and lipids [9, 68, 70]. Nevertheless, the high instrumental costs and the requirement for highly specialized personnel remain key limitations to its widespread adoption [7].

IR spectroscopy is based on the absorption of light, being classified according to the applied spectral range into near-infrared (NIR), mid-infrared (MIR), and far-infrared (FIR) regions [67]. The NIR (800–2500 nm; 12,500–4000 cm^{-1}) is associated with overtone and combination bands resulting from molecular vibrations induced by light-molecule interactions. Although these bands are broad, they contain valuable qualitative and quantitative information that can be extracted using chemometric tools. The MIR range (2500–25,000 nm; 4000–400 cm^{-1}) encompasses fundamental absorption bands that form a molecular "fingerprint," which is widely used for structural characterization and spectral comparison. This region is richer in sharper and more intense bands than NIR, making interpretation easier, while also enabling quantitative analyses. The FIR range (25,000–100,000 nm; 400–10 cm^{-1}) is less accessible and primarily used for studying inorganic molecules, and consequently, it has limited relevance for cannabinoid analysis [3].

IR spectroscopy offers an attractive alternative to conventional techniques, especially in applications demanding rapid results and in-field analysis, such as forensic applications [71]. Once a chemometric model has been developed and integrated into the device, IR analysis becomes simple, rapid, non-invasive, and user-friendly, without the need for highly qualified personnel. However, the development of a robust chemometric model remains a critical step, demanding the expertise of highly qualified professionals to ensure accurate calibration and reliable performance [3].

Raman spectroscopy is based on the phenomenon of light scattering, in which monochromatic radiation interacts with a sample [67]. A small fraction of the scattered photons (approximately 1 in 10 million photons) is inelastically scattered, carrying information about the vibrational modes of chemical species present. This vibrational "fingerprint" makes Raman spectroscopy a highly selective technique [72]. Like IR, Raman spectrophotometers can also offer rapidity, greenness, and the possibility of performing in-field analysis, in addition to resulting in spectra usually easier to interpret than NIR spectra and do not demand complex algorithms to treat the chemical information [3]. However, the main limitation of these techniques relies on the negative effect of fluorescence and the lack of sensitivity [67].

Because of their complementary structural information, IR and Raman spectroscopies are often used together. Both techniques share several advantages: they require little or no sample preparation, are non-invasive, do not need the use of organic solvents, and offer rapid analysis times. These features translate into notable time and cost savings while making the methods eco-friendlier. A key distinction between the two techniques lies in volume of sample interrogated. For instance, in MIR spectroscopy with an attenuated total reflection (ATR) interface, the light penetration depth is typically 0.5–2 μm, depending on the crystal and the wavenumber. In contrast, NIR and Raman spectroscopy have a deeper penetration—approximately 1–3 mm—depending on the sample nature, the wavelength, and the light power applied [67].

Recent applications of these techniques include the routine monitoring of Δ^9-THC in cannabis plant and resins using NIR spectroscopy [73]. NIR has also been used to quantify eight different cannabinoids in ground leaves and inflorescences from cannabis, in addition to the discrimination between illegal and legal cannabis varieties [74, 75]. Beyond plant materials, NIR has demonstrated sensitivity and specificity for quantifying CBD in different liquid pharmaceutical products, underscoring its potential for real-time monitoring of CBD content during production processes [76]. Additionally, a handheld Raman spectrometer combined with orthogonal partial least squares was successfully applied to develop a classification model capable of discriminating Δ^9-THC-rich, CBD-rich cannabis plants and hemp [77]. Portable vibrational spectroscopy instruments are highly suitable for continuous monitoring of the main cannabinoids in all growth stages of cannabis, as well as during the production of cannabis-derived products [9].

References

1. Duchateau C, Stévigny C, Waeytens J, Deconinck E (2025) Chromatographic and spectroscopic analyses of cannabinoids: a narrative review focused on cannabis herbs and oily products. Molecules 30. https://doi.org/10.3390/molecules30030490
2. Giese MW, Lewis MA, Giese L, Smith KM (2015) Development and validation of a reliable and robust method for the analysis of cannabinoids and terpenes in cannabis. J AOAC Int 98:1503–1522. https://doi.org/10.5740/jaoacint.15-116

3. Deidda R, Dispas A, De Bleye C, Hubert P, Ziemons É (2022) Critical review on recent trends in cannabinoid determination on cannabis herbal samples: from chromatographic to vibrational spectroscopic techniques. Anal Chim Acta 1209:339184. https://doi.org/10.1016/J.ACA.2021.339184

4. Citti C, Braghiroli D, Vandelli MA, Cannazza G (2018) Pharmaceutical and biomedical analysis of cannabinoids: a critical review. J Pharm Biomed Anal 147:565–579. https://doi.org/10.1016/j.jpba.2017.06.003

5. Welling MT, Liu L, Hazekamp A, Dowell A, King GJ (2019) Developing robust standardised analytical procedures for cannabinoid quantification: laying the foundations for an emerging cannabis-based pharmaceutical industry. Med Cannabis Cannabinoids 2:1–13. https://doi.org/10.1159/000496868

6. Oliveira IGC, Grecco CF, de Souza ID, Queiroz MEC (2024) Current chromatographic methods to determine cannabinoids in biological samples: a review of the state-of-the art on sample preparation techniques. Green Analytical Chemistry 11:100161. https://doi.org/10.1016/J.GREEAC.2024.100161

7. Inamassu CH, Raspini e Silva L, Marchioni C (2024) Recent advances in the chromatographic analysis of endocannabinoids and phytocannabinoids in biological samples. J Chromatogr A 1732:465225. https://doi.org/10.1016/J.CHROMA.2024.465225

8. Battista N, Sergi M, Montesano C, Napoletano S, Compagnone D, Maccarrone M (2014) Analytical approaches for the determination of phytocannabinoids and endocannabinoids in human matrices. Drug Test Anal 6:7–16. https://doi.org/10.1002/dta.1574

9. Stefkov G, Karanfilova IC, Gjorgievska VS, Trajkovska A, Geskovski N, Karapandzova M, Kulevanova S (2022) Analytical techniques for phytocannabinoid profiling of cannabis and cannabis-based products—a comprehensive review. Molecules 27. https://doi.org/10.3390/molecules27030975

10. Fairbairn JW, Liebmann JA (1973) The extraction and estimation of the cannabinoids in *Cannabis sativa* L. and its products. J Pharm Pharmacol 25:150–155. https://doi.org/10.1111/j.2042-7158.1973.tb10609.x

11. UNODC (2022) Recommended methods for the identification and analysis of cannabis and cannabis products, Vienna

12. Jin D, Dai K, Xie Z, Chen J (2020) Secondary metabolites profiled in cannabis inflorescences, leaves, stem barks, and roots for medicinal purposes. Sci Rep 10. https://doi.org/10.1038/s41598-020-60172-6.

13. American Herbal Pharmacopoeia (2014) Cannabis inflorescence—standards of identity, analysis, and quality control, Scotts Valley

14. Gunjević V, Grillo G, Carnaroglio D, Binello A, Barge A, Cravotto G (2021) Selective recovery of terpenes, polyphenols and cannabinoids from Cannabis sativa L. inflorescences under microwaves. Ind Crops Prod 162:113247. https://doi.org/10.1016/J.INDCROP.2021.113247

15. DAB (2018) Geschäftsstelle der Arzneibuch-Kommissionen. Bundesinstitut fur Arzneimittel und Miedizinprodukte, Monografie Cannabis blüten, Bonn

16. Hazekamp A, Simons R, Peltenburg-Looman A, Sengers M, Van Zweden R, Verpoorte R (2004) Preparative isolation of cannabinoids from Cannabis sativa by centrifugal partition chromatography. J Liq Chromatogr Relat Technol 27:2421–2439. https://doi.org/10.1081/JLC-200028170

17. Wianowska D, Dawidowicz AL, Kowalczyk M (2015) Transformations of Tetrahydrocannabinol, tetrahydrocannabinolic acid and cannabinol during their extraction from Cannabis sativa L. J Anal Chem 70:920–925. https://doi.org/10.1134/S1061934815080183

18. Raharjo TJ, Verpoorte R (2004) Methods for the analysis of cannabinoids in biological materials: a review. Phytochem Anal 15:79–94. https://doi.org/10.1002/pca.753

19. Silva EMP, Vitiello A, Miro A, Ribeiro CJA (2025) Recent HPLC-UV approaches for cannabinoid analysis: from extraction to method validation and quantification compliance. Pharmaceuticals 18:786. https://doi.org/10.3390/ph18060786

20. Yang S, Sun M (2024) Recent advanced methods for extracting and analyzing cannabinoids from cannabis-infused edibles and detecting hemp-derived contaminants in food (2013–2023):

a comprehensive review. J Agric Food Chem 72:13476–13499. https://doi.org/10.1021/acs. jafc.4c01286

21. Ştefănescu R, Vlad RA, Imre S, Tero-Vescan A, Ősz BE, Sita DD, Farczádi L (2025) Development and validation of an LC-MS/MS method for the quantification of six cannabinoids in commercial products. Studia Universitatis Babeş-Bolyai Chemia 70:173–190. https://doi.org/ 10.24193/subbchem.2025.1.12

22. Ciolino LA, Ranieri TL, Taylor AM (2018) Commercial cannabis consumer products part 1: GC–MS qualitative analysis of cannabis cannabinoids. Forensic Sci Int 289:429–437. https:// doi.org/10.1016/J.FORSCIINT.2018.05.032

23. Minamoto K, Takayama T, Katehashi H, Katagi M, Inoue K (2024) Development and validation of a sensitive and simultaneous liquid chromatography tandem mass spectrometry method for the determination of eight phytocannabinoids in various CBD products. J Pharm Biomed Anal 249:116341. https://doi.org/10.1016/J.JPBA.2024.116341

24. Lindsay CM, Abel WD, Jones-Edwards EE, Brown PD, Bernard KK, Taylor TT (2021) Form and content of Jamaican cannabis edibles. J Cannabis Res 3:29. https://doi.org/10.1186/s42 238-021-00079-9

25. Christinat N, Savoy MC, Mottier P (2020) Development, validation and application of a LC-MS/MS method for quantification of 15 cannabinoids in food. Food Chem 318:126469. https:// doi.org/10.1016/J.FOODCHEM.2020.126469

26. Perestrelo R, Silva P, Porto-Figueira P, Pereira JAM, Silva C, Medina S, Câmara JS (2019) QuEChERS—fundamentals, relevant improvements, applications and future trends. Anal Chim Acta 1070:1–28. https://doi.org/10.1016/J.ACA.2019.02.036

27. Kole PL, Venkatesh G, Kotecha J, Sheshala R (2011) Recent advances in sample preparation techniques for effective bioanalytical methods. Biomed Chromatogr 25:199–217. https://doi. org/10.1002/bmc.1560

28. Boyaci E, Rodríguez-Lafuente Á, Gorynski K, Mirnaghi F, Souza-Silva ÉA, Hein D, Pawliszyn J (2015) Sample preparation with solid phase microextraction and exhaustive extraction approaches: comparison for challenging cases. Anal Chim Acta 873:14–30. https://doi.org/ 10.1016/J.ACA.2014.12.051

29. Khatibi SA, Hamidi S, Siahi-Shadbad MR (2022) Application of liquid-liquid extraction for the determination of antibiotics in the foodstuff: recent trends and developments. Crit Rev Anal Chem 52:327–342. https://doi.org/10.1080/10408347.2020.1798211

30. Shah I, Al-Dabbagh B, Salem AE, Hamid SAA, Muhammad N, Naughton DP (2019) A review of bioanalytical techniques for evaluation of cannabis (Marijuana, weed, Hashish) in human hair. BMC Chem 13:106. https://doi.org/10.1186/s13065-019-0627-2

31. Bian Y, Zhang Y, Zhou Y, Wei B, Feng X (2023) Recent insights into sample pretreatment methods for mycotoxins in different food matrices: a critical review on novel materials. Toxins (Basel) 15:215. https://doi.org/10.3390/toxins15030215

32. Li W, Jian W, Fu Y (2019) Basic sample preparation techniques in LC-MS bioanalysis. In: Sample preparation in LC-MS bioanalysis, Wiley, pp 1–30. https://doi.org/10.1002/978111927 4315.ch1

33. Cruz JC, Miranda LFC, Queiroz MEC (2021) Pipette tip micro-solid phase extraction (octyl-functionalized hybrid silica monolith) and ultra-high-performance liquid chromatography-tandem mass spectrometry to determine cannabidiol and tetrahydrocannabinol in plasma samples. J Sep Sci 44:1621–1632. https://doi.org/10.1002/jssc.202000906

34. Jain R, Singh R (2016) Microextraction techniques for analysis of cannabinoids. TrAC, Trends Anal Chem 80:156–166. https://doi.org/10.1016/J.TRAC.2016.03.012

35. Madden O, Walshe J, Kishore Patnala P, Barron J, Meaney C, Murray P (2023) Phytocannabinoids—an overview of the analytical methodologies for detection and quantification of therapeutically and recreationally relevant cannabis compounds. Crit Rev Anal Chem 53:211–231. https://doi.org/10.1080/10408347.2021.1949694

36. Tran J, Elkins AC, Spangenberg GC, Rochfort SJ (2022) High-throughput quantitation of cannabinoids by liquid chromatography triple-quadrupole mass spectrometry. Molecules 27:742. https://doi.org/10.3390/molecules27030742

37. Hazekamp A, Peltenburg A, Verpoorte R, Giroud C (2005) Chromatographic and spectroscopic data of cannabinoids from *Cannabis sativa* L. J Liq Chromatogr Relat Technol 28:2361–2382. https://doi.org/10.1080/10826070500187558

38. Fiorini D, Molle A, Nabissi M, Santini G, Benelli G, Maggi F (2019) Valorizing industrial hemp (Cannabis sativa L.) by-products: cannabidiol enrichment in the inflorescence essential oil optimizing sample pre-treatment prior to distillation. Ind Crops Prod 128:581–589. https://doi.org/10.1016/j.indcrop.2018.10.045

39. Hart ED, Vikingsson S, Mitchell JM, Winecker RE, Flegel R, Hayes ED (2022) Conversion of 7-carboxy-cannabidiol (7-COOH-CBD) to 11-Nor-9-carboxy-tetrahydrocannabinol (THC-COOH) during sample preparation for GC–MS analysis. J Anal Toxicol 46:573–576. https://doi.org/10.1093/jat/bkab046

40. Andrews R, Paterson S (2012) Production of identical retention times and mass spectra for Δ9-tetrahydrocannabinol and cannabidiol following derivatization with trifluoracetic anhydride with 1,1,1,3,3,3-hexafluoroisopropanol*. J Anal Toxicol 36:61–65. https://doi.org/10.1093/jat/bkr017

41. Rosendo LM, Rosado T, Oliveira P, Simão AY, Margalho C, Costa S, Passarinha LA, Barroso M, Gallardo E (2022) The determination of cannabinoids in urine samples using microextraction by packed sorbent and gas chromatography-mass spectrometry. Molecules 27:5503. https://doi.org/10.3390/molecules27175503

42. Pourseyed Lazarjani M, Torres S, Hooker T, Fowlie C, Young O, Seyfoddin A (2020) Methods for quantification of cannabinoids: a narrative review. J Cannabis Res 2:35. https://doi.org/10.1186/s42238-020-00040-2

43. Leghissa A, Hildenbrand ZL, Schug KA (2018) A review of methods for the chemical characterization of cannabis natural products. J Sep Sci 41:398–415. https://doi.org/10.1002/jssc.201701003

44. Tremlová B, Mikulášková HK, Hajduchová K, Jancikova S, Kaczorová D, Ćavar Zeljković S, Dordevic D (2021) Influence of technological maturity on the secondary metabolites of hemp concentrate (Cannabis sativa L.). Foods 10:1418. https://doi.org/10.3390/foods10061418

45. Glivar T, Eržen J, Kreft S, Zagožen M, Čerenak A, Čeh B, Tavčar Benković E (2020) Cannabinoid content in industrial hemp (Cannabis sativa L.) varieties grown in Slovenia. Ind Crops Prod 145:112082. https://doi.org/10.1016/J.INDCROP.2019.112082

46. Ilias Y, Rudaz S, Mathieu P, Veuthey J-L, Christen P (2004) Analysis of cannabis material by headspace solid-phase microextraction combined with gas chromatography-mass spectrometry. Chimia (Aarau) 58:219. https://doi.org/10.2533/000942904777677957

47. Cardenia V, Gallina Toschi T, Scappini S, Rubino RC, Rodriguez-Estrada MT (2018) Development and validation of a fast gas chromatography/mass spectrometry method for the determination of cannabinoids in cannabis sativa L. J Food Drug Anal 26:1283–1292. https://doi.org/10.1016/J.JFDA.2018.06.001

48. Amirav A, Neumark B, Margolin Eren KJ, Fialkov AB, Tal N (2021) Cannabis and its cannabinoids analysis by gas chromatography–mass spectrometry with Cold EI. J Mass Spectrom 56. https://doi.org/10.1002/jms.4726

49. Aizpurua-Olaizola O, Soydaner U, Öztürk E, Schibano D, Simsir Y, Navarro P, Etxebarria N, Usobiaga A (2016) Evolution of the cannabinoid and terpene content during the growth of *cannabis sativa* plants from different chemotypes. J Nat Prod 79:324–331. https://doi.org/10.1021/acs.jnatprod.5b00949

50. Aizpurua-Olaizola O, Omar J, Navarro P, Olivares M, Etxebarria N, Usobiaga A (2014) Identification and quantification of cannabinoids in cannabis sativa L. plants by high performance liquid chromatography-mass spectrometry. Anal Bioanal Chem 406:7549–7560. https://doi.org/10.1007/s00216-014-8177-x

51. Andrews R, Paterson S (2012) A validated method for the analysis of cannabinoids in post-mortem blood using liquid–liquid extraction and two-dimensional gas chromatography–mass spectrometry. Forensic Sci Int 222:111–117. https://doi.org/10.1016/J.FORSCIINT.2012.05.007

52. Nahar L, Chaiwut P, Sangthong S, Theansungnoen T, Sarker SD (2024) Progress in the analysis of phytocannabinoids by HPLC and UPLC (or UHPLC) during 2020–2023. Phytochem Anal 35:927–989. https://doi.org/10.1002/pca.3374

53. Omar J, Olivares M, Amigo JM, Etxebarria N (2014) Resolution of co-eluting compounds of cannabis sativa in comprehensive two-dimensional gas chromatography/mass spectrometry detection with multivariate curve resolution-alternating least squares. Talanta 121:273–280. https://doi.org/10.1016/J.TALANTA.2013.12.044

54. D'Atri V, Fekete S, Clarke A, Veuthey J-L, Guillarme D (2019) Recent advances in chromatography for pharmaceutical analysis. Anal Chem 91:210–239. https://doi.org/10.1021/acs.analchem.8b05026

55. Micalizzi G, Vento F, Alibrando F, Donnarumma D, Dugo P, Mondello L (2021) Cannabis Sativa L.: a comprehensive review on the analytical methodologies for cannabinoids and terpenes characterization. J Chromatogr A 1637:461864. https://doi.org/10.1016/j.chroma.2020.461864

56. Tzimas PS, Petrakis EA, Halabalaki M, Skaltsounis LA (2021) Effective determination of the principal non-psychoactive cannabinoids in fiber-type Cannabis sativa L. by UPLC-PDA following a comprehensive design and optimization of extraction methodology. Anal Chim Acta 1150:338200. https://doi.org/10.1016/J.ACA.2021.338200

57. Durante C, Anceschi L, Brighenti V, Caroli C, Afezolli C, Marchetti A, Cocchi M, Salamone S, Pollastro F, Pellati F (2022) Application of experimental design in HPLC method optimisation for the simultaneous determination of multiple bioactive cannabinoids. J Pharm Biomed Anal 221:115037. https://doi.org/10.1016/J.JPBA.2022.115037

58. Citti C, Ciccarella G, Braghiroli D, Parenti C, Vandelli MA, Cannazza G (2016) Medicinal cannabis: principal cannabinoids concentration and their stability evaluated by a high performance liquid chromatography coupled to diode array and quadrupole time of flight mass spectrometry method. J Pharm Biomed Anal 128:201–209. https://doi.org/10.1016/j.jpba.2016.05.033

59. Speybrouck D, Lipka E (2016) Preparative supercritical fluid chromatography: a powerful tool for chiral separations. J Chromatogr A 1467:33–55. https://doi.org/10.1016/j.chroma.2016.07.050

60. Tarafder A (2016) Metamorphosis of supercritical fluid chromatography to SFC: an overview. TrAC, Trends Anal Chem 81:3–10. https://doi.org/10.1016/j.trac.2016.01.002

61. Wang M, Wang Y, Avula B, Radwan MM, Wanas AS, Mehmedic Z, van Antwerp J, ElSohly MA, Khan IA (2017) Quantitative determination of cannabinoids in cannabis and cannabis products using ultra-high-performance supercritical fluid chromatography and diode array/mass spectrometric detection. J Forensic Sci 62:602–611. https://doi.org/10.1111/1556-4029.13341

62. Wang M, Wang Y-H, Avula B, Radwan MM, Wanas AS, van Antwerp J, Parcher JF, ElSohly MA, Khan IA (2016) Decarboxylation study of acidic cannabinoids: a novel approach using ultra-high-performance supercritical fluid chromatography/photodiode array-mass spectrometry. Cannabis Cannabinoid Res 1:262–271. https://doi.org/10.1089/can.2016.0020

63. Breitenbach S, Rowe WF, McCord B, Lurie IS (2016) Assessment of ultra high performance supercritical fluid chromatography as a separation technique for the analysis of seized drugs: applicability to synthetic cannabinoids. J Chromatogr A 1440:201–211. https://doi.org/10.1016/J.CHROMA.2016.02.047

64. Jambo H, Dispas A, Avohou HT, André S, Hubert C, Lebrun P, Ziemons É, Hubert P (2018) Implementation of a generic SFC-MS method for the quality control of potentially counterfeited medicinal cannabis with synthetic cannabinoids. J Chromatogr B 1092:332–342. https://doi.org/10.1016/J.JCHROMB.2018.05.049

65. Felletti S, De Luca C, Buratti A, Bozza D, Cerrato A, Capriotti AL, Laganà A, Cavazzini A, Catani M (2021) Potency testing of cannabinoids by liquid and supercritical fluid chromatography: where we are, what we need. J Chromatogr A 1651:462304. https://doi.org/10.1016/j.chroma.2021.462304

66. Guillarme D, Desfontaine V, Heinisch S, Veuthey J-L (2018) What are the current solutions for interfacing supercritical fluid chromatography and mass spectrometry? J Chromatogr B 1083:160–170. https://doi.org/10.1016/j.jchromb.2018.03.010

67. Deidda R, Sacre P-Y, Clavaud M, Coïc L, Avohou H, Hubert P, Ziemons E (2019) Vibrational spectroscopy in analysis of pharmaceuticals: critical review of innovative portable and handheld NIR and Raman spectrophotometers. TrAC, Trends Anal Chem 114:251–259. https://doi.org/10.1016/j.trac.2019.02.035

68. Politi M, Peschel W, Wilson N, Zloh M, Prieto JM, Heinrich M (2008) Direct NMR analysis of cannabis water extracts and tinctures and semi-quantitative data on Δ9-THC and Δ9-THC-acid. Phytochemistry 69:562–570. https://doi.org/10.1016/J.PHYTOCHEM.2007.07.018

69. Happyana N, Agnolet S, Muntendam R, Van Dam A, Schneider B, Kayser O (2013) Analysis of cannabinoids in laser-microdissected trichomes of medicinal Cannabis sativa using LCMS and cryogenic NMR. Phytochemistry 87:51–59. https://doi.org/10.1016/J.PHYTOCHEM.2012.11.001

70. Hazekamp A, Choi YH, Verpoorte R (2004) Quantitative analysis of cannabinoids from cannabis sativa using 1H-NMR. Chem Pharm Bull (Tokyo) 52:718–721. https://doi.org/10.1248/cpb.52.718

71. de Araujo WR, Cardoso TMG, da Rocha RG, Santana MHP, Muñoz RAA, Richter EM, Paixão TRLC, Coltro WKT (2018) Portable analytical platforms for forensic chemistry: a review. Anal Chim Acta 1034:1–21. https://doi.org/10.1016/j.aca.2018.06.014

72. Dumont E, De Bleye C, Sacré P-Y, Netchacovitch L, Hubert P, Ziemons E (2016) From near-infrared and Raman to surface-enhanced Raman spectroscopy: progress. Limit Perspect Bioanal Bioanal 8:1077–1103. https://doi.org/10.4155/bio-2015-0030

73. Deidda R, Coppey F, Damergi D, Schelling C, Coïc L, Veuthey J-L, Sacré P-Y, De Bleye C, Hubert P, Esseiva P, Ziemons É (2021) New perspective for the in-field analysis of cannabis samples using handheld near-infrared spectroscopy: a case study focusing on the determination of Δ9-tetrahydrocannabinol. J Pharm Biomed Anal 202:114150. https://doi.org/10.1016/j.jpba.2021.114150

74. Sánchez-Carnerero Callado C, Núñez-Sánchez N, Casano S, Ferreiro-Vera C (2018) The potential of near infrared spectroscopy to estimate the content of cannabinoids in Cannabis sativa L.: a comparative study. Talanta 190:147–157. https://doi.org/10.1016/J.TALANTA.2018.07.085

75. Duchateau C, Kauffmann J, Canfyn M, Stévigny C, De Braekeleer K, Deconinck E (2020) Discrimination of legal and illegal Cannabis spp. according to European legislation using near infrared spectroscopy and chemometrics. Drug Test Anal 12:1309–1319. https://doi.org/10.1002/dta.2865

76. Espel Grekopoulos J (2019) Construction and validation of quantification methods for determining the cannabidiol content in liquid pharma-grade formulations by means of near-infrared spectroscopy and partial least squares regression. Med Cannabis Cannabinoids 2:43–55. https://doi.org/10.1159/000500266

77. Sanchez L, Baltensperger D, Kurouski D (2020) Raman-based differentiation of hemp, cannabidiol-rich hemp, and cannabis. Anal Chem 92:7733–7737. https://doi.org/10.1021/acs.analchem.0c00828

Chapter 5
Practical Aspects for the Incorporation of Cannabinoids in Food Products

5.1 Cannabinoids Sources

Cannabis-derived food products are often referred to as functional foods not only for their nutritional value but also for the health benefits that the cannabinoids, terpenes, essential fatty acids, and other phytonutrients can offer. Interest in functional foods has increased over the last decades and development of cannabis-added food products became a veritable race [1].

Currently, the most popular cannabis-derived edible products containing cannabinoids are produced from seeds, hemp flour, and extracts [2]. The cannabinoids are primarily produced and accumulated in the trichomes, which are mainly concentrated in the flowers of female plants.

5.1.1 Hemp Seeds

Hemp seeds, which can be processed as a whole, reduced to flour, or transformed into oil, are reported to contain only trace amounts of cannabinoids, with Δ^9-THC content far below the legal limits of 0.3% established in United States and Europe [3]. However, the cannabinoids composition varies largely among different cannabis varieties and their concentration in the final products must be accurately determined [3]. In spite of that, hemp seeds are considered a valuable source of nutrients for functional foods, mainly ascribed to its proteins (20–25%) and lipids (25–35%) contents [4]. Whole hemp seed proteins are considered easily digestible due to the absence of protease inhibitors and contain adequate amount of essential amino acids, making hemp seeds an interesting source of vegetable protein [5]. Regarding the fatty acids profile, hemp seed's oil contains up to 90% of unsaturated fatty acids, of which 70–80% is composed by polyunsaturated fatty acids and are characterized by optimal ω-6/ω-3 ratio of 3:1 [3, 6].

© The Author(s), under exclusive license to Springer Nature Switzerland AG 2026
S. E. Azam et al., *Cannabinoids in Food Science*,
Chemistry of Foods, https://doi.org/10.1007/978-3-032-11546-1_5

In addition to the interesting protein and lipidic profile of hemp seeds, the whole seed contains around 30% of carbohydrates, of which about 6% are water soluble and 22% are insoluble polysaccharides [7]. Despite this substantial carbohydrate content, the nature and composition of the non-cellulosic polysaccharides fraction, including xylan, xyloglucan, and pectin, remain relatively understudied [3]. These polysaccharides contribute to dietary fiber, which is essential for human nutrition, as it corresponds to the edible portion of plant material that resists enzymatic digestion in the small intestine and undergoes partial or complete fermentation in the large intestine [4].

Other minor compounds such as vitamins and minerals are also present in hemp seeds [8]. The most prevalent are the fat-soluble vitamin E (tocopherols), which help mitigate oxidative stress, and vitamin A (mostly β-carotene), essential for healthy skin development [8]. The mineral content of hemp seeds ranges from approximately 4 to 7%, although its minerals profile has been relatively under-explored [3]. Authors have reported that the mineral composition of hemp seeds is affected by environmental conditions, soil mineral composition, and plant variety [8].

5.1.2 Hemp Extracts

Although the cannabinoids are found in both hemp and marijuana varieties of cannabis, this section focuses on hemp extracts as the primary source, since hemp and its derivatives with low Δ^9-THC content (less than 0.2% according to European legislation) are not considered a controlled substance. Consequently, hemp extracts are the most widely used in the production of cannabis-based food products. To meet regulatory requirements, it is essential to ensure that the cannabinoids source has a consistent and reproducible composition across batches, which in turns demands the implementation of rigorous cultivation, harvesting, and processing practices to provide a reliable supply [9].

In addition, producing cannabis products from reliable sources is crucial to ensure accurate cannabinoids dosing, particularly for products marketed for their functional benefits [10]. For instance, in THC-containing products, precise dose control is critical for consumer safety, as small variations in the ingested amount (e.g., between 5 and 10 mg) can lead to markedly different effects. In contrast, for CBD-added products, dosing accuracy is more closely linked to consumer trust than safety, since CBD is well tolerated, good safety profile, and exhibits no potential for abuse or dependence [10].

When concentrated extracts are used as cannabinoids source, the extraction processes—whether or not followed or not by purification—become a crucial step to enable their incorporation in various products [11, 12]. Extracts containing Δ^9-THC and/or CBD usually dissolve in edible oils, but their lipophilic nature requires the use of specific carriers to enable their incorporation in aqueous-based products [1, 9, 11]. Additional challenges for the formulation of cannabis edibles include the inherently bitter taste of the cannabis extracts, which requires the use of flavoring agents to make the final product more appealing to consumers [13], and their viscous and sticky nature, which requires preheating to 70–80 °C to increase fluidity and facilitate handling [1]. The main strategies to facilitate the incorporation of cannabinoids in food matrices include encapsulation and/or emulsification of cannabis extracts [1]. These delivery systems typically consist of a non-polar core, where cannabinoids are solubilized, surrounded by a polar shell that interacts with the aqueous environment. Examples include microemulsions, nanoemulsions, emulsions, biopolymer nanoparticles, solid lipid nanoparticles, and filled hydrogels beads [9]. The production methods of these delivery systems as well as their advantages and disadvantages are discussed below.

5.2 Food Matrices

Among the cannabinoid-added food products, the beverages (e.g., tea, coffee, fruit juices, soft drinks, beers, wines, and liquors), baked goods (e.g., cookies, brownies, and cakes), and confectionery products (chocolates, caramels, gummies) are the most commonly found [11]. Each of them has a different composition and chemical structure, which influence the bioavailability and bioactivity of the cannabinoids [9].

Beverages are primarily aqueous-based products that, due to their nature, typically contain relatively low concentrations of cannabinoids, usually 5–10 mg per serving [9]. To enable their incorporation into aqueous beverages, the cannabinoids can be chemically modified by attaching a hydrophilic moiety, such as a carbohydrate group, which increases water solubility and allow direct dispersion into the beverages or other food products [14]. However, cannabinoids are more commonly entrapped into delivery systems before being incorporated into aqueous-based products. In some cases, the beverages are converted into a powder form, usually by spray drying, which can be redispersed in water by the consumer, or the consumer can add the cannabinoids to the beverages using a tincture—cannabis oil dispersed in alcohol [9].

Another popular category of cannabis foods is the baked goods. These products usually contain sugar, flour, fat, eggs, proteins, and possibly salt and flavors. After mixing all together, they are heated in oven to cook. In these products, the cannabinoids tend to dissolve in the fat phase, usually constituted of butter, margarine, or shortening [9]. The cannabinoids present boiling point at temperature around 150–180 °C, which can make them sensitive to vaporization during baking process, usually occurring at temperatures around 180–210 °C. During baking, the cannabinoids can undergo chemical transformations such as decarboxylation of the acidic forms to its active neutral form, or even degradation. For example, Δ^9-THC can be converted into CBN at temperatures around 85 °C [1, 13].

Confectionery products constitute one of the main delivery forms of cannabinoids. In chocolate-based products containining cocoa butter, cannabinoids are mainly located in the fatty domains, which may enhance their bioavailability by promoting the formation of mixed micelles and chylomicrons within the gastrointestinal tract

[9]. In contrast, gummies are soft chewy gels, traditionally made from gelatin or other plant-based gelling agents, such as pectin, agar, and carrageenan. As these products do not contain a fat source, the incorporation of cannabinoids within them is more difficult, and their prior addition into carrier systems is necessary [9].

5.3 Stability of Cannabinoids During Food Processing

The cannabinoids added to food matrices are prone to chemical transformation due to heat, oxidation, and interaction with other ingredients, such as enzymatic-induced transformation [13]. However, data relating the stability of individual cannabinoids is still limited, with most of the literature focusing on the stability of Δ^9-THC and CBD [15]. The stability of cannabinoids in cannabis plant material and extracts during storage conditions, namely, under refrigeration and ambient temperatures, with and without light exposure are available in the literature [15], but cannabinoids stability in food matrices under processing conditions still requires further investigation.

The stability of cannabinoids in hemp leaves and inflorescences stored under different conditions—at 5 °C in the dark and at 20 °C with and without light—showed that Δ^9-THC is highly susceptible to temperature and light-induced degradation [16]. It was observed that storage at 5 °C in the dark led to a 10% decrease in Δ^9-THC content, while the storage at 20 °C in the dark caused 25% reduction. The most drastic condition of storage at 20 °C with light exposure resulted in a reduction of 63% in the Δ^9-THC content [16]. The same authors also evaluated the stability of powdered cannabis resins packed and stored at 20 °C with and without light. Samples collected from the surface presented a reduction in the Δ^9-THC content from 11.6 to 5.2%, whereas samples collected 2 mm below the surface and from the middle showed a final content of 11.4%. Another study evaluating the stability of Δ^9-THC, CBD, and CBN under various conditions (in dark at -20 °C, in light without air exposure at room temperature, in light with air exposure at room temperature, and in dark without light exposure at room temperature) observed that the samples exposed to air, whether in dark or light, presented the largest Δ^9-THC losses, evidencing the critical role of oxygen in its degradation [17]. This study further demonstrated that CBD was more stable than Δ^9-THC regardless of storage conditions. Considering the lipidic nature of cannabinoids, their susceptibility to oxidative degradation upon oxygen exposure is expected [18].

Similarly, hemp seed oil is prone to oxidation and rancidity due to its high content of unsaturated fatty acids, such as α-linoleic acid and γ-linolenic acid [15, 19]. Since hemp seed oil is typically obtained by cold pressing, it predominantly contains cannabinoids in their acidic forms, such as CBDA and THCA. Therefore, evaluating the ratio between CBDA and CBD can serve as a useful indicator of storage conditions, as prolonged storage or improper conditions may lead to conversion of CBDA into CBD [18].

Decarboxylation is a non-enzymatic reaction only influenced by temperature, therefore the higher the temperature, the faster the decarboxylation [20]. However, current knowledge about the stability of cannabinoids under heating conditions remains limited, making it difficult to establish their safety profile [18]. A study demonstrated that Δ^9-THCA was completely converted into Δ^9-THC after heating at 160 °C for 15 min, while CBN was additionally formed at 180 °C [21]. Another study evaluated the decarboxylation of hemp inflorescences before extraction at temperatures in the range of 85–145 °C, concluding that 115 °C was the optimal temperature to achieve a complete conversion of Δ^9-THCA to Δ^9-THC without the formation of CBN after 40 min [22]. Peschel [23]. evaluated the influence of pasteurization conditions (70 °C for 2 h and 80 °C for 20 min) on the cannabinoids content of ethanol-based tinctures and reported a 20% reduction in the total cannabinoids content (Δ^9-THCA, Δ^9-THC, and CBN). However, these pasteurization conditions did not accelerate the decarboxylation of Δ^9-THCA. Studies regarding the stability of cannabinoids in food matrices demonstrate that it is highly dependent on the composition of the matrix as well as the processing and storage conditions [15]. However, a recent study demonstrated that the process used to prepare and bake a brownie did not affect the Δ^9-THC and CBD contents [24]. Another study evaluated the effects of baking temperature and time on the cannabinoids profile of hemp seed breads, demonstrating that both parameters had significant effect cannabinoids stability, promoting decarboxylation and a change of Δ^9-THC content by -7 to $+53\%$ [25]. Further systematic research on the cannabinoid stability in different food matrices and storage processes is urgently needed, as the degradation compounds can represent a prospective public health risk [26]. Given the stability profile of the cannabinoids in different food matrices, formulations must be optimized to guarantee accurate delivery of the declared cannabinoid content and to ensure shelf-life of the cannabis product [13].

5.4   Trends to Increase Solubility and Bioavailability of Cannabinoids in Food Products

Recent trends for the development of cannabis-based products focus on new alternatives to increase the water solubility and bioavailability of cannabinoids [9, 27, 28]. The inherently limited water solubility of cannabinoids represents a major challenge for formulating aqueous-based foods products and contributes to their low oral bioavailability. Furthermore, extensive first-pass hepatic metabolism further reduces the concentration of cannabinoids reaching the systemic circulation. Consequently, the design of innovative approaches to overcome these limitations has become a central focus of current research. Promising approaches include the development of excipient foods and delivery systems, which are further detailed in this section.

5.4.1 Co-administration and Excipient Foods

It has been reported that food affects the bioavailability of hydrophobic compounds, including cannabinoids [9, 29–31]; however, systematic studies describing the food-matrix effects on the pharmacokinetic of cannabinoids in humans are still limited. A study reporting the pharmacokinetic profile of Δ^9-THC in humans after ingestion of cannabis-loaded brownies with different doses (10, 25 and 50 mg Δ^9-THC) demonstrated that the maximum blood levels (C_{max}) were relatively low for all concentrations evaluated (1–3.5 ng/mL) compared to the levels obtained after smoking cannabis (15–192 ng/mL) [32]. In addition, the brownies took much longer (2–3 h) to reach the peak concentration in blood (t_{max}) than smoking (less than 10 min). For CBD, it has been reported that the bioavailability of CBD is significantly higher when it is ingested with a meal [31]. Actually, the co-administration of CBD with a high-fat meal increases its bioavailability by fourfold [30]. This effect is mainly ascribed to the impact of the ingested lipids on the intestinal bioaccessibility and lymphatic transport of the cannabinoids [33].

However, it is important to consider that the chain length of the lipids co-ingested with cannabinoids can affect the micellization and uptake mechanisms in the small intestine [34]. While medium-chain triglycerides (MCT)—chain length of 6–12 carbons—are rapidly taken up into the enterocyte and enter systemic circulation through portal vein, long-chain triglycerides (LCT), which has chain lengths ranging from 13 to 21 carbons, require micellization to be then packed into chylomicrons before entering the lymphatic system [34]. The compounds absorbed via portal vein are processed in the liver and undergo phase I and phase II metabolism for complete or partial deactivation before being released into the bloodstream, which is not the case for the compounds taken via lymphatic system [35]. These findings highlight the potential for development of new strategies to improve the cannabinoids bioavailability by manipulating dietary parameters.

Another very interesting approach to control the cannabinoids bioavailability is the co-administration of natural absorption enhancers, also called bioenhancers [36]. These compounds are agents which themselves do not possess inherent pharmacological activity of their own but when co-administered with an active compound enhances its bioavailability and efficacy [37]. Some alkaloids, flavonoids, and several phenolic compounds act as bioenhancers by inhibiting the action of major drug-metabolizing enzymes of the CYP-450 family, P-glycoprotein efflux pumps, and UDP-glucuronyl-transferases [37–39]

Cherniakov et al. [36] explored the effect of three potential bioenhancers—piperine, curcumin, and resveratrol—on the oral bioavailability of CBD and Δ^9-THC in rats. These compounds were co-administered with the cannabinoids using a self-emulsifying drug delivery system, named as Pro NanoLipospheres (PNL). Among the tested bioenhancers, piperine showed the most pronounced effect: CBD-piperine-PNL increased the area under the curve (AUC) by sixfold compared to CBD solution, while THC-piperine-PNL led to a 9.3-fold increase in AUC compared to THC solution. The same research group advanced the study to evaluate the performance of

PNL system delivering both cannabinoids CBD and Δ^9-THC with piperine (CBD/Δ^9-THC-piperine-PNL) in a clinical setting [40]. They observed that the CBD/Δ^9-THC-piperine-PNL resulted in a threefold increase in C_{max} and 1–fivefold increase in AUC for Δ^9-THC when compared to the commercial formulation Sativex®, while for CBD, a fourfold increase in C_{max} and 2.2-fold increase in AUC were observed.

These findings raise the potential of the development of excipient foods, a concept recently introduced by McClements [41, 42] as an alternative to increase the bioavailability of hydrophobic bioactive compounds, such as cannabinoids, in foods, drugs, and supplements. An excipient food does not contain cannabinoids itself; rather, it is carefully designed with ingredients or structures that, while have no bioactivity themselves, can improve the cannabinoids bioavailability when co-ingested with it [42].

For instance, one strategy is the formulation of excipient foods containing lipid-based components that, once ingested, rapidly form mixed micelles capable of dissolving the cannabinoids and transporting them to the intestinal cells where they are absorbed [9]. Additionally, other ingredients, such as bioenhancers, can be included in excipient foods to modulate the metabolism and absorption of the cannabinoids. A good candidate for use as an excipient food should be desirable for the consumers and suitable for consumption in a daily basis to ensure good compliance, in addition to presenting sufficiently long shelf-life to do not require frequent purchases [42]. For example, a potential excipient food could be a fluid shot or candy that would be taken before consuming the cannabis-based food to modulate its pharmacokinetic profile and biological effects [9].

However, McClements and Xiao [42] advertise about the technical, legal, and commercial challenges related with the development and commercialization of excipient foods, which are largely due to the difficulty of proving their efficacy in enhancing nutraceutical bioactivity. Unlike pharmaceuticals, which can be administered in precise doses under controlled conditions, nutraceuticals are typically consumed at low levels within complex diets, making it hard to establish clear correlations between intake, bioavailability, and health outcomes. This complicates the ability of food manufacturers to generate the robust evidence required by regulators to support specific health claims, thereby reducing industry incentives for investment. Additional hurdles include variability in consumer diets, the timing and type of excipient–nutraceutical co-ingestion, the need for diverse food matrices to suit different nutraceuticals and consumer preferences, and the potential for adverse health effects if excipients inadvertently increase the absorption of harmful compounds or raise beneficial compounds to toxic levels. Therefore, beyond scientific validation, successful commercialization will also depend on careful risk–benefit communication to consumers, with clear guidance on which excipient foods should be paired with specific nutraceutical-rich products.

5.4.2   Lipid-Based Delivery Systems

A wide variety of lipid-based delivery systems have been explored to encapsulate cannabinoids, offering promising strategies to address the major challenges in formulating cannabis-based food products, particularly their poor water solubility and limited bioavailability. Each of the different platforms that have been used to vehiculate cannabinoids present distinct characteristics and advantages. In general, these systems differ in terms of ingredients and manufacturing processes required to produce them, which in turn also impacts their robustness in different food matrices, as well as the final appearance, texture, and flavor [9]. Furthermore, they also vary in terms of their potential impact on the bioavailability and pharmacokinetic profiles of cannabinoids after ingestion [9].

5.4.2.1   Nanoemulsions and Microemulsions

Nanoemulsions and microemulsions are colloidal dispersions typically formulated from oil, water, and a surfactant, which are widely used to encapsulate hydrophobic components to enable their dispersion into aqueous medium [43]. Although nanoemulsions and microemulsions share many structural similarities, there are important differences between them [43]. Microemulsions are spontaneously formed by mixing a surfactant with water, often in presence of additional components, such as oil, co-surfactants, or co-solvents, typically resulting in the smallest particles sizes among colloidal delivery systems (around 5–50 nm) [9]. Structurally, they consist of a hydrophobic core formed by the surfactant tails and a hydrophilic shell formed by their polar heads, enabling efficient solubilization of cannabinoids. Their main advantages as delivery systems include thermodynamic stability, which extends shelf-life, and ease production, since they form spontaneously once the production conditions (temperature and composition) are optimized without the need for specialized equipment [9]. However, their formulation requires high surfactant concentrations, which can impart undesirable taste, increase toxicity risk, and raise productions costs. In addition, their small hydrophobic core limits cannabinoid loading capacity [9]. Owing to these constraints and the nature of the components typically employed, microemulsions are more commonly applied in pharmaceutical contexts than in food products. Actually, Lazzari et al. [44]. developed a microemulsion with very small particle size (6.7–8.0 nm) by only mixing a surfactant (Solutol®HS15) and Δ^9-THC in water.

Unlike microemulsions, nanoemulsions are thermodynamically unstable colloidal dispersions composed of two immiscible phases—oil and water—stabilized by emulsifiers that require the input of external energy to convert the separate components into a colloidal dispersion with particles sizes usually between 100 and 500 nm [43, 45]. Common methods to produce nanoemulsions comprise high-energy approaches, such as ultrasonication or high-pressure homogenization [46]. Although microemulsions and nanoemulsions can be produced from exactly the same ingredients, the microemulsions typically need greater surfactant-to-oil ratio [43]. Nanoemulsions, on the other hand, offer important advantages, including a higher loading capacity that allows greater incorporation of hydrophobic active compounds, as well as the possibility of being formulated with natural emulsifiers instead of synthetic ones, which are frequently associated with adverse health effects [47]. Nanoemulsions are a versatile colloidal system for incorporation of cannabinoids into food systems. Actually, numerous commercial products have been created using nanoemulsions to protect the cannabinoids from degradation during storage, as well as to achieve specific pharmacokinetic profiles after ingestion, however, academic studies in this area still limited, mainly due to legal restrictions worldwide [9, 45].

Nakano et al. [48] developed a nanoemulsion constituted of vitamin E acetate (1.7%, w/w), ethanol (3.8%, w/w), Tween 20 (70%, w/w), and distilled water (24.5%, w/w) to evaluate the oral absorption of CBD in rats. The formulation, containing 30 mg CBD/mL with mean particle size of 35.3 ± 11.8 nm, was able to increase the AUC by 65% and reduce the T_{max} from 8.0 to 2.4 h, in comparison with a control CBD oil.

Other studies have evaluated the use of natural emulsifiers, such as Q-naturale® (containing 15% of *Quillaja* saponin) in nanoemulsions with CBD extracts [49, 50]. A formulation with 10% of soybean oil, 10% of Q-naturale® emulsifier, and 16.6% of CBD extract produced nanoemulsions with average particle diameter of 120 nm, showing good stability for 6 weeks under ambient conditions [49]. In another study, olive oil (7.2%) combined with 4.5% of Q-naturale® and 5.4% of CBD extract resulted in nanoemulsions with particle size (d_{50}) of 110 nm [50]. More recently, cholinium oleate, a biocompatible ionic liquid, was investigated as an emulsifier to produce isolate CBD nanoemulsions [28]. Formulations containing 10% of hemp seed oil and 1% of CBD were prepared with either 1% or 2% of ionic liquid. The 1% ionic liquid formulation presented a particle diameter of 225 nm and remained stable for 4 weeks at ambient conditions.

5.4.2.2   Liposomes

Liposomes are spherical vesicles, typically in the range from 50 nm to 50 μm, constituted of a hydrophilic core enclosed by one or more phospholipid bilayers [51]. Whitin these structures, hydrophobic compounds, such as cannabinoids, are entrapped in the non-polar regions formed by the phospholipid tails of the bilayer [9]. The structural properties of liposomes, particularly their rigidity and physical stability, are strongly influenced by the type of phospholipid used [52]. For example, liposomes prepared with phosphatidylcholines form more permeable and less stable bilayers, whereas saturated phospholipids with long acyl chains provide the opposite effect, enhancing stability [53]. However, the inherent poor stability of liposomes remains a major limitation for their use as delivery system in food matrices [9]. To overcome this, cholesterol is often incorporated into the phospholipid bilayers to improve membrane rigidity and stability [9]. The addition of biopolymeric coating is also a strategy used to increase the liposomes stability [54]. The incorporation of cannabinoids into the liposomal bilayer also influences its stability in different ways: CBD and CBN have been shown to increase the molecular order rigidity of the bilayer, whereas Δ^9-THC has the opposite effect, reducing rigidity [52]. Another disadvantages of liposomes as delivery system for cannabinoids include their low encapsulation efficiency, which results from their limited ability to hold and stabilize the compounds within the phospholipid bilayer and their short release time [54].

Liposomes can be produced by a range of methods varying in cost, time, ease, and equipment needs, and the selection of the most suitable fabrication method depends on the type of liposomes, phospholipid properties, as well as the interactions between the phospholipids and the dispersion medium [9, 54]. The liposomes production methods include detergent depletion, membrane extrusion, lipid layer hydration, high-pressure homogenization, ethanol injection method, sonication, reverse-phase evaporation, and microfluidization [54].

Although liposomes can be used to increase the oral bioavailability of hydrophobic compounds, very few studies are available in the literature of their application for cannabinoids. A liposomal formulation based on sunflower lecithin (phosphatidyl-choline) was developed to improve the CBD oral bioavailability [55]. The results from the assay with health human volunteers demonstrated that the oral bioavailability of CBD from liposomes was 17-fold higher than that of free CBD at 1-h post-administration.

5.4.2.3   Solid-Lipid Nanoparticles and Nanostructured Lipid Carriers

Solid-lipid nanoparticles (SLNs) and nanostructured lipid carriers (NLCs) have increased in popularity for the delivery of cannabinoids, particularly CBD and Δ^9-THC [35]. NLCs are considered the second generation of SLNs, essentially developed to overcome their limitations [54]. NLCs contain a combination of solid and liquid lipids, in contract to SLNs that contain only solid lipids. The presence of liquid lipids results in structural imperfections or less ordered crystalline arrangement, thus offering higher loading capacity and a more controlled release of the bioactive compounds [46, 52]. Another interesting advantage of NLCs is their capacity to increase the bioavailability of hydrophobic compounds [54]. Due to these attractive characteristics, NLCs have been chosen as delivery systems for cannabinoids, mainly in pharmaceutical applications [56–59].

Hommoss et al. [57] and Matarazzo et al. [56] developed mucoadhesive NLCs for nasal delivery of Δ^9-THC and CBD, respectively. Δ^9-THC-loaded NLCs were prepared using hot high-pressure homogenization technique, combining an aqueous phase of water with cetylpyridinium chloride (CPC) as stabilizer and a lipid phase of cetyl palmitate and Δ^9-THC [57]. Tween 80 was also evaluated as a co-stabilizer. Two optimal formulations achieved particles sizes around 200 nm: one containing 1% particles (lipid: Δ^9-THC ratio 7:1) stabilized with 0.05% CPC, and another containing 2% particles stabilized with 0.05% CPC and 0.05% Tween 80. Both systems provided effective stabilization of Δ^9-THC and promoted rapid drug release, supporting enhanced nasal absorption [57]. A CBD-loaded NLC was produced by hot microemulsion technique using stearic acid (1.25%) as solid lipid, oleic acid (0.75%) as a liquid lipid, CPC (0.05%) as active surfactant, and Span 20® (0.25%) as co-surfactant [56]. These particles had a mean diameter of 177 nm, a high entrapment efficiency (99.99%), and a positive surface charge (+41 mV). They presented a biphasic release profile, with an initial burst releasing over 50% of CBD within 5 min, followed by a slower and sustained release [56].

Recently, NLCs formulated only using food-grade ingredients were developed to incorporate both CBD extract and isolate CBD [27, 28]. In these formulations, the lipid phase (10%) consisted of hemp seed oil as liquid lipid and fully hydrogenated soybean oil as solid lipid. One system employed a 60:40 (w/w) ratio of liquid-to-solid lipids with 3% of soy lecithin as emulsifier [27] while another used a 50:50 (w/w) ratio stabilized with 1 of % ionic liquid as emulsifier [28]. Both formulations were prepared with 1% CBD, either as extract or isolate. These food-grade NLCs presented particles sizes around 230–270 nm, with good stability over 4 weeks of storage at ambient conditions. In addition, in vitro digestion results indicated that NLCs provided higher CBD stability during gastrointestinal digestion [28]. These findings emphasize the potential of NLCs as suitable delivery systems for cannabinoids in nutraceutical and functional foods.

5.4.2.4 Self-emulsifying Drug Delivery Systems

Another promising approach for the encapsulation of cannabinoids is the use of self-emulsifying drug delivery systems (SEDDS), also referred to as micro-(SMEDDS) or nano-(SNEDDS) formulations, depending on the resulting particle size, stability, and bioavailability, which varies significantly according to the ingredients used in their composition [60]. Unlike microemulsions and nanoemulsions, which must be pre-formed before administration, SEDDS are emulsion pre-concentrates in which the lipophilic bioactive compound is mixed with lipids, surfactants, and co-solvents. Upon contact with gastrointestinal fluids in the upper lumen, they spontaneously emulsify, enhancing both solubility and oral absorption of the encapsulated molecules [46, 52].

In SEDDS, surfactants are essential components, because they reduce the surface tension at the oil–water interface and stabilize the emulsion [61]. Surfactants and co-surfactants typically represent the largest portion of the formulation, accounting for approximately 30–60% of the system [62]. However, high surfactant concentrations may cause gastrointestinal irritation and alter the host microbiota, underscoring the importance of carefully optimizing excipients levels to minimize potential toxicity [60].

Although several scientific studies have reported the use of SEDDS for cannabinoids delivery in pharmaceutical applications [36, 40, 63–68], none have explored their development for food applications using food-grade ingredients. For instance, as previously mentioned, Cherniakov et al. [36, 40] developed advanced SEDDS formulations incorporating bioenhancers to improve the biopharmaceutical properties of cannabinoids. The optimized formulation was prepared by mixing ethyl acetate and lecithin at a 4:1 ratio, followed by the addition of triglyceride tricaprin, polyoxyl-40-hydroxyl castor oil, Tween 20, and Span 80 in a 1:1:1:1 ratio. After gently stirring and heating at 40 °C, Δ^9-THC, CBD and piperine were added to the pre-concentrate and gently stirred until a homogenous solution was obtained [36].

Similarly, De Prá et al. [63] developed an SEDDS formulation comprising glycerol monolinoleate (Maisine® CC), polyoxyl-40 castor oil, polyethylene glycol 400, and CBD extract, to enhance the oral absorption of CBD. The resulting system exhibited particle sizes in the range of 148–194 nm. In mice, this SEDDS increased the AUC of CBD by 3.97-fold compared to a conventional CBD in MCT oil, while in humans it achieved a 1.48-fold increase. Another SEDDS formulation based on the VESIsorb® technology constituted of emulsifiers, edible vegetable oils, and fatty acids yields was prepared to load CBD extract yielding particle sizes of 40–50 nm [65]. Following single oral administration in humans, SEDDS formulation increased the AUC of CBD by 2.85-fold compared to the reference formulation constituted of CBD in MCT oil.

As demonstrated, SEDDS are predominantly formulated as liquids and are most commonly delivered in soft or hard gelatin capsules for oral administration [60]. Despite their advantages, liquid SEDDS present notable limitations, including poor physical and chemical stability, and restricted applicability in solid dosage forms, which can hinder large-scale production and compromise patient adherence [69] To address these challenges, recent research has increasingly focused on solid SEDDS, which preserves the bioavailability benefits of liquid systems while offering superior stability, portability, and versatility [60]. Solid SEDDS can be easily produced using stablished techniques such as spray drying, melt extrusion, or adsorption onto inert carriers [62]. Given these combined advantages, solid SEDDS emerge as a particularly promising strategy for the incorporation of cannabinoids into diverse food matrices. A recent study developed a Δ^9-THC/CBD self-emulsifying powder, which is a SNEEDS-based formulation constituted of 43% sucrose, 29% maltodextrin, 16% cannabis oil (providing 8 mg Δ^9-THC and 8 mg CBD per 500 mg of formulation), 9% olive oil, and 3% glycyrrhizic acid (ammonium salt) as a natural emulsifier [68]. The manufacturing process involved pre-emulsification, high-pressure nano-emulsification, and freeze-drying with a matrix of disaccharides and polysaccharides. The resulting powder exhibited an average particle size of 157 nm. Pharmacokinetic evaluation in humans demonstrated that this formulation more than doubled the relative bioavailability of cannabinoids compared to conventional oil drops, indicating enhanced absorption and a faster onset of action [68].

5.5 Public Health Concerns and Regulatory Status

Regulatory frameworks for cannabis-based products are under continuous debate and revision, with a growing trend toward legalization worldwide: This shift has driven a crescent expansion of cannabis markets. Within this scenario, cannabis edible products are gaining popularity, particularly among young consumers [70].

After widespread prohibition of cannabis cultivation in the early twentieth century motivated by the psychotropic effects of Δ^9-THC, policies began to change in recent decades [11]. Only by the end of twentieth century, hemp varieties with low Δ^9-THC content were re-approved for cultivation as an industrial crop in the European Union (EU). This was made possible under the Common Agricultural Policy (CAP), which permitted the cultivation of hemp varieties listed in the EU Common Catalogue of Varieties of Agricultural Plant Species, provided their Δ^9-THC content did not exceed 0.3% [71]. In the United States, hemp cultivation policy changed in 2018, when hemp varieties with less than 0.3% Δ^9-THC were removed from the Drug Enforcement Administration's list of controlled substances [5].

Regulations governing cannabinoids content in food products remain highly heterogeneous worldwide, with restrictions most often targeting Δ^9-THC [15]. Some countries set maximum Δ^9-THC limits for specific food categories, while others establish limits for "total Δ^9-THC," which comprises the sum of Δ^9-THC and its acidic precursor Δ^9-THCA [2]. This approach is based on the fact that Δ^9-THCA can readily decarboxylate into Δ^9-THC during heating processes such as cooking or baking, thereby altering the final Δ^9-THC concentration in the products and potentially pose a risk to consumers [2]. Despite differences across countries, the established limits for Δ^9-THC in foods are consistently in the mg/kg range [2].

Hemp seeds and derivatives (oil, protein, flour) are already used as food ingredients. In United States, they are classified as Generally Recognized As Safe (GRAS) by Food and Drugs Administration (FDA), while in the EU they are considered as "not novel." However, only trace amounts of cannabinoids are permitted in these ingredients [2]. In contrast, other plant parts and their extracts containing cannabinoids, particularly CBD, are classified as novel foods under EU Regulation EU2015/2283 [72]. According to this regulation, a novel food is defined as any food that has not been consumed to a significant degree by humans within EU before May 2007 [72], CBD-containing products must undergo a safety assessment by European Food Safety Authority (EFSA) and obtain approval from European Commission before being placed on the market.

Despite this requirement, CBD is not classified as controlled substance in the EU, and products containing CBD are therefore not explicitly regulated. This regulatory gap has led to a wide range of CBD products being marketed without mandatory novel food authorization [15, 73, 74]. A similar situation exists in the United States, where FDA has not yet established a clear regulatory framework for CBD. As a result, CBD-containing products are widely available on the market without FDA approval, despite safety concerns raised by the agency [12].

Cannabis companies often employ aggressive marketing strategies to promote these products, sometimes misleading the consumers. A recent study evaluated over two thousand cannabis edible products from U.S. dispensaries, analyzing product labeling, child-oriented features, health and non-health claims, warnings, and online promotional strategies. Results showed that more than 80% of products contained at least 100 mg of Δ^9-THC, and over 20% displayed child-appealing elements, factors that increase the risk of overconsumption and accidental ingestion [70]. Even more critical is the issue of inaccurate labeling of cannabis products [15].

Although producers are expected to ensure proper labeling, several studies have shown that the cannabinoids content in many commercial products does not match the values declared by the manufacturers [74–76], highlighting the potential health risks associated with the lack of control over cannabis edibles.

To better understand the origin of adverse effects associated with the consumption of CBD products, Lachenmeier et al. [68] investigated two possible hypotheses: (i) conversion of CBD to Δ^9-THC in stomach under acidic conditions and (ii) contamination with residual Δ^9-THC due to the use crude hemp extracts containing full spectrum to formulate CBD products. Their results discharged the first hypothesis, as assays with pure CBD solutions in simulated gastric fluids or under various storage conditions (e.g., heat and light) did not result in Δ^9-THC formation. However, 25% of the tested products from the German market (mostly CBD oils) contained Δ^9-THC levels above the lowest observed adverse effects level (2.5 mg/day) while CBD concentrations were below the no observed adverse effects level. These findings suggest that some adverse effects attributed to CBD products may, in fact, result from Δ^9-THC contamination. Furthermore, all samples tested in this study presented non-compliance with EU food law, including unsafe Δ^9-THC levels, use of full-spectrum hemp extracts without novel food approvals, unauthorized heath claims, and deficiencies in mandatory food labeling requirements [74].

To address these risks, some authorities set recommendations for acceptable daily intake (ADI) of cannabinoids. The UK Food Standards Agency (FSA) has proposed an ADI of 10 mg/day of CBD (0.15 mg/kg for a 70 kg adult) and an upper safe limit for Δ^9-THC of 0.07 mg Δ^9-THC/day (equivalent to 1 µg/kg/day for a 70 kg adult) [77]. This Δ^9-THC limit aligns with the acute human reference dose of 1 µg Δ^9-THC/kg established by EFSA [78]. Similarly, the Swiss Federal Food Safety and Veterinary Office has set a limit of 12 mg/day of CBD for a 70 kg adult [12].

Overall, the safety, efficacy, and purity of cannabis products on the market remains questionable. There is an urgent need for further studies on chemical transformations cannabinoids undergo during storage to guarantee that the declared concentrations are maintained during the entire shelf-life. Equally important is the development of effective regulations able to control safety and quality of commercialized products [15, 74, 79].

5.6 Future Perspectives

The rediscovery of the therapeutic potential of cannabinoids, supported by recent scientific findings and facilitated by continuous changing in the regulatory restrictions of their medical and recreational use, has motivated a surge of commercial interest in cannabis-based products. Among these, edibles represent one of the fastest growing categories, with market expansion projected to continue in the near future. Despite this promising scenario, the incorporation of cannabinoids into food products faces several scientific, technological, and regulatory challenges that must be addressed to ensure safe, effective, and consistent products.

One of the most critical issues is the stability of cannabinoids during storage and food processing. Cannabinoids are highly sensitive to factors such as temperature, light, oxygen, and pH, which considerably affects their stability in the products during processing and storage, causing their degradation, ultimately leading to the formation of undesirable compounds and significant losses in the expected bioactivity. However, systematic studies evaluating the kinetics of cannabinoid degradation across different processing technologies and storage conditions are still limited. Scientific development in this area is urgently needed to enable design of robust food matrices, technological processes, and packaging systems capable of protecting cannabinoids and ensuring product stability through shelf-life. These aspects also raise the necessity of advanced analytical techniques to enable the identification and monitoring the content of cannabinoids and, eventually, their metabolites across different food matrices, since it has been demonstrated that constituents of the food matrices influence the analyses results.

Another major challenge is the limited understanding of the metabolic pathways that cannabinoids undergo following oral ingestion, particularly the influence of the food matrix and host physiology on their absorption, distribution, metabolism, and excretion. While studies have advanced the knowledge on the CBD and Δ^9-THC metabolism, many other cannabinoids and their metabolites remain poorly characterized, restricting the ability to predict their bioactivity and long-term health effects in food applications. The establishment of accurate knowledge about the metabolism of cannabinoids is required to further evolve the dose recommendations for the cannabinoids oral intake in order to guarantee the expected efficacy without observing side effects related to their ingestion.

From a technological perspective, future trends in the field of cannabinoids in food products will likely focus on improving the bioaccessibility and bioavailability of cannabinoids, which are constrained by their high lipophilicity and poor water solubility. As previously discussed, innovative formulation strategies are emerging, mainly based on lipidic carriers, which enhance micellar incorporation during digestion and promote effective epithelial transport. In addition to these features, there is still space for future strategies that can also reduce the intra- and interindividual variability in response to oral ingestion of cannabinoids, influenced by genetic factors and health status, which currently represents a challenge for designing standardized food products. The development of hybrid systems that combine natural surfactants, polysaccharides, or proteins with lipid carriers may represent a promising next generation of cannabinoid delivery in food products.

Another area requiring further exploitation is the interactions of cannabinoids with other dietary components and the gut microbiota. These interactions may modulate cannabinoids stability, metabolism, and therapeutic efficacy, but remain insufficiently understood. The role of the ingestion of lipids on the increase of cannabinoids bioavailability, at least for CBD and Δ^9-THC, has been demonstrated, but the possible interactions of cannabinoids with other constituents are still scarce in literature.

On the regulatory side, discrepancies across countries and regions continue to hinder the global development of cannabinoid-based foods. The classification of cannabinoids as novel foods in EU and the lack of harmonized safety assessments worldwide create barriers for product development and commercialization. This lack of regulation on the products containing cannabinoids results in a plethora of products being illegally marketed containing inaccurate concentration of cannabinoids with respect to the declared values, which represents a potential risk to the public health. Several studies have been demonstrating that this a persistent issue worldwide. Future progress will depend on the establishment of clearer regulatory frameworks, standardized safety testing, and evidence-based health claims to ensure consumer protection while fostering innovation.

Overall, future developments of cannabinoid-containing food products will likely combine advanced formulation strategies to overcome bioavailability barriers with a deeper scientific understanding of cannabinoid metabolism and interactions within the human body. At the same time, regulatory harmonization will be essential to translating these advances into safe, effective, and widely accessible products.

Acknowledgements This work was supported by the Portuguese Foundation for Science and Technology (FCT) under the scope of the strategic funding of UID/04469: Centre of Biological Engineering of the University of Minho, and by LABBELS—Associate Laboratory in Biotechnology, Bioengineering and Microelectromechanical Systems, LA/P/0029/2020. Shahnshah E. Azam thanks FCT for his fellowship (reference 2024.06762.BD).

References

1. Marangoni IP, Marangoni AG (2019) Cannabis edibles: dosing, encapsulation, and stability considerations. Curr Opin Food Sci 28:1–6. https://doi.org/10.1016/j.cofs.2019.01.005
2. Christinat N, Savoy MC, Mottier P (2020) Development, validation and application of a LC-MS/MS method for quantification of 15 cannabinoids in food. Food Chem 318:126469. https://doi.org/10.1016/J.FOODCHEM.2020.126469
3. Sirangelo TM, Diretto G, Fiore A, Felletti S, Chenet T, Catani M, Spadafora ND (2025) Nutrients and bioactive compounds from cannabis sativa seeds: a review focused on omics-based investigations. Int J Mol Sci 26:5219. https://doi.org/10.3390/ijms26115219
4. Tănase Apetroaei V, Pricop EM, Istrati DI, Vizireanu C (2024) Hemp seeds (Cannabis sativa L.) as a valuable source of natural ingredients for functional foods—a review. Molecules 29:2097. https://doi.org/10.3390/molecules29092097
5. Rizzo G, Storz MA, Calapai G (2023) The role of hemp (Cannabis sativa L.) as a functional food in vegetarian nutrition. Foods 12:3505. https://doi.org/10.3390/foods12183505
6. Farinon B, Molinari R, Costantini L, Merendino N (2020) The seed of industrial hemp (Cannabis sativa L.): nutritional quality and potential functionality for human health and nutrition. Nutrients 12:1935. https://doi.org/10.3390/nu12071935
7. Julakanti S, Charles APR, Syed R, Bullock F, Wu Y (2023) Hempseed polysaccharide (Cannabis sativa L.): physicochemical characterization and comparison with flaxseed polysaccharide. Food Hydrocoll 143:108900. https://doi.org/10.1016/J.FOODHYD.2023.108900
8. Montero L, Ballesteros-Vivas D, Gonzalez-Barrios AF, Sánchez-Camargo ADP (2023) Hemp seeds: nutritional value, associated bioactivities and the potential food applications in the Colombian context. Front Nutr 9. https://doi.org/10.3389/fnut.2022.1039180
9. McClements DJ (2020) Enhancing efficacy, performance, and reliability of cannabis edibles: insights from lipid bioavailability studies. Annu Rev Food Sci Technol:45–70. https://doi.org/10.1146/annurev-food-032519

10. Malochleb M (2019) Why Cannabis edibles are creating a buzz. Food Technol Mag 73:32–44
11. Rasera GB, Ohara A, de Castro RJS (2021) Innovative and emerging applications of cannabis in food and beverage products: from an illicit drug to a potential ingredient for health promotion. Trends Food Sci Technol 115:31–41. https://doi.org/10.1016/J.TIFS.2021.06.035
12. Knutson K (2020) Food safety lessons for cannabis-infused edibles. Elsevier. https://doi.org/10.1016/C2019-0-00117-3
13. King JW (2019) The relationship between cannabis/hemp use in foods and processing methodology. Curr Opin Food Sci 28:32–40. https://doi.org/10.1016/j.cofs.2019.04.007
14. Akhtar MT, Mustafa NR, Verpoorte R (2015) Hydroxylation and glycosylation of Δ^9-etrahydrocannabinol by *Catharanthus roseus* cell suspension culture. Biocatal Biotransform 33:279–286. https://doi.org/10.3109/10242422.2016.1151006
15. Kanabus J, Bryła M, Roszko M, Modrzewska M, Pierzgalski A (2021) Cannabinoids—characteristics and potential for use in food production. Molecules 26:6723. https://doi.org/10.3390/molecules26216723
16. Fairbairn JW, Liebmann JA, Rowan MG (1976) The stability of cannabis and its preparations on storage. J Pharm Pharmacol 28:1–7. https://doi.org/10.1111/j.2042-7158.1976.tb04014.x
17. Grafström K, Andersson K, Pettersson N, Dalgaard J, Dunne SJ (2019) Effects of long term storage on secondary metabolite profiles of cannabis resin. Forensic Sci Int 301:331–340. https://doi.org/10.1016/J.FORSCIINT.2019.05.035
18. Bartončíková M, Lapčíková B, Lapčík L, Valenta T (2023) Hemp-derived CBD used in food and food supplements. Molecules 28:8047. https://doi.org/10.3390/molecules28248047
19. Wang M, Wang Y, Avula B, Radwan MM, Wanas AS, Mehmedic Z, van Antwerp J, ElSohly MA, Khan IA (2017) Quantitative determination of cannabinoids in cannabis and cannabis products using ultra-high-performance supercritical fluid chromatography and diode array/mass spectrometric detection. J Forensic Sci 62:602–611. https://doi.org/10.1111/1556-4029.13341
20. Citti C, Pacchetti B, Vandelli MA, Forni F, Cannazza G (2018) Analysis of cannabinoids in commercial hemp seed oil and decarboxylation kinetics studies of cannabidiolic acid (CBDA). J Pharm Biomed Anal 149:532–540. https://doi.org/10.1016/J.JPBA.2017.11.044
21. Dussy FE, Hamberg C, Luginbühl M, Schwerzmann T, Briellmann TA (2005) Isolation of Δ9-THCA-A from hemp and analytical aspects concerning the determination of Δ9-THC in cannabis products. Forensic Sci Int 149:3–10. https://doi.org/10.1016/J.FORSCIINT.2004.05.015
22. Casiraghi A, Roda G, Casagni E, Cristina C, Musazzi U, Franzè S, Rocco P, Giuliani C, Fico G, Minghetti P, Gambaro V (2018) Extraction method and analysis of cannabinoids in cannabis olive oil preparations. Planta Med 84:242–249. https://doi.org/10.1055/s-0043-123074
23. Peschel W (2016) Quality control of traditional cannabis tinctures: pattern, markers, and stability. Sci Pharm 84:567–584. https://doi.org/10.3390/scipharm84030567
24. Wolf CE, Poklis JL, Poklis A (2016) Stability of tetrahydrocannabinol and cannabidiol in prepared quality control medible brownies. J Anal Toxicol. https://doi.org/10.1093/jat/bkw114
25. Lindekamp N, Triesch N, Weiss M, Voß A, Rautenberg T, Rohn S, Weigel S (2025) Hemp containing breads: cannabinoid content, profiles, and thermal stability during baking. Food Chem 487:144714. https://doi.org/10.1016/J.FOODCHEM.2025.144714
26. AL Ubeed HMS, Brennan CS, Schanknecht E, Alsherbiny MA, Saifullah M, Nguyen K, Vuong QV (2022) Potential applications of hemp (*Cannabis sativa* L.) extracts and their phytochemicals as functional ingredients in food and medicinal supplements: a narrative review. Int J Food Sci Technol 57:7542–7555. https://doi.org/10.1111/ijfs.16116
27. Vardanega R, Lüdtke FL, Loureiro L, Gonçalves RFS, Pinheiro AC, Vicente AA (2024) Development and characterization of nanostructured lipid carriers for cannabidiol delivery. Food Chem 441. https://doi.org/10.1016/j.foodchem.2023.138295
28. Vardanega R, Lüdtke FL, Loureiro L, Toledo Hijo AAC, Martins JT, Pinheiro AC, Vicente AA (2024) Enhancing cannabidiol bioaccessibility using ionic liquid as emulsifier to produce nanosystems: characterization of structures, cytotoxicity assessment, and in vitro digestion. Food Res Int 188. https://doi.org/10.1016/j.foodres.2024.114498

29. Sitovs A, Logviss K, Lauberte L, Mohylyuk V (2024) Oral delivery of cannabidiol: revealing the formulation and absorption challenges. J Drug Deliv Sci Technol 92:105316. https://doi. org/10.1016/J.JDDST.2023.105316

30. Perucca E, Bialer M (2020) Critical aspects affecting cannabidiol oral bioavailability and metabolic elimination, and related clinical implications. CNS Drugs 34:795–800. https://doi. org/10.1007/s40263-020-00741-5

31. Silmore LH, Willmer AR, Capparelli EV, Rosania GR (2021) Food effects on the formulation, dosing, and administration of cannabidiol (CBD) in humans: a systematic review of clinical studies. Pharmacother J Hum Pharmacol Drug Ther 41:405–420. https://doi.org/10.1002/phar. 2512

32. Vandrey R, Herrmann ES, Mitchell JM, Bigelow GE, Flegel R, LoDico C, Cone EJ (2017) Pharmacokinetic profile of oral cannabis in humans: blood and oral fluid disposition and relation to pharmacodynamic outcomes. J Anal Toxicol 41:83–99. https://doi.org/10.1093/jat/bkx012

33. Zgair A, Lee JB, Wong JCM, Taha DA, Aram J, Di Virgilio D, McArthur JW, Cheng Y-K, Hennig IM, Barrett DA, Fischer PM, Constantinescu CS, Gershkovich P (2017) Oral administration of cannabis with lipids leads to high levels of cannabinoids in the intestinal lymphatic system and prominent immunomodulation. Sci Rep 7:14542. https://doi.org/10.1038/s41598-017-15026-z

34. Izgelov D, Shmoeli E, Domb AJ, Hoffman A (2020) The effect of medium chain and long chain triglycerides incorporated in self-nano emulsifying drug delivery systems on oral absorption of cannabinoids in rats. Int J Pharm 580. https://doi.org/10.1016/j.ijpharm.2020.119201.

35. Light K, Karboune S (2022) Emulsion, hydrogel and emulgel systems and novel applications in cannabinoid delivery: a review. Crit Rev Food Sci Nutr 62:8199–8229. https://doi.org/10. 1080/10408398.2021.1926903

36. Cherniakov I, Izgelov D, Domb AJ, Hoffman A (2017) The effect of Pro NanoLipospheres (PNL) formulation containing natural absorption enhancers on the oral bioavailability of delta-9-tetrahydrocannabinol (THC) and cannabidiol (CBD) in a rat model. Eur J Pharm Sci 109:21–30. https://doi.org/10.1016/j.ejps.2017.07.003

37. Ajazuddin A, Alexander A, Qureshi L, Kumari P, Vaishnav M, Sharma S, Saraf S (2014) Saraf, role of herbal bioactives as a potential bioavailability enhancer for active pharmaceutical ingredients. Fitoterapia 97:1–14. https://doi.org/10.1016/J.FITOTE.2014.05.005

38. Bhardwaj RK, Glaeser H, Becquemont L, Klotz U, Gupta SK, Fromm MF (2002) Piperine, a major constituent of black pepper, inhibits human P-glycoprotein and CYP3A4. J Pharmacol Exp Ther 302:645–650. https://doi.org/10.1124/JPET.102.034728

39. Detampel P, Beck M, Krähenbühl S, Huwyler J (2012) Drug interaction potential of resveratrol. Drug Metab Rev 44:253–265. https://doi.org/10.3109/03602532.2012.700715

40. Cherniakov I, Izgelov D, Barasch D, Davidson E, Domb AJ, Hoffman A (2017) Piperine-pro-nanolipospheres as a novel oral delivery system of cannabinoids: pharmacokinetic evaluation in healthy volunteers in comparison to buccal spray administration. J Control Release 266:1–7. https://doi.org/10.1016/j.jconrel.2017.09.011

41. McClements DJ, Zou L, Zhang R, Salvia-Trujillo L, Kumosani T, Xiao H (2015) Enhancing nutraceutical performance using excipient foods: designing food structures and compositions to increase bioavailability. Compr Rev Food Sci Food Saf 14:824–847. https://doi.org/10.1111/ 1541-4337.12170

42. McClements DJ, Xiao H (2014) Excipient foods: designing food matrices that improve the oral bioavailability of pharmaceuticals and nutraceuticals. Food Funct 5:1320–1333. https:// doi.org/10.1039/C4FO00100A

43. McClements DJ (2012) Nanoemulsions versus microemulsions: terminology, differences, and similarities. Soft Matter 8:1719–1729. https://doi.org/10.1039/C2SM06903B

44. Lazzari P, Fadda P, Marchese G, Casu GL, Pani L (2010) Antinociceptive activity of Δ9-tetrahydrocannabinol non-ionic microemulsions. Int J Pharm 393:239–244. https://doi.org/10. 1016/J.IJPHARM.2010.04.010

45. McClements DJ (2021) Advances in edible nanoemulsions: digestion, bioavailability, and potential toxicity. Prog Lipid Res 81:101081. https://doi.org/10.1016/J.PLIPRES.2020.101081

46. Reddy TS, Zomer R, Mantri N (2023) Nanoformulations as a strategy to overcome the delivery limitations of cannabinoids. Phytother Res 37:1526–1538. https://doi.org/10.1002/ptr.7742

47. Dammak I, do PJ, Sobral A, Aquino A, das Neves MA, Conte-Junior CA (2020) Nanoemulsions: using emulsifiers from natural sources replacing synthetic ones—a review. Compr Rev Food Sci Food Saf 19:2721–2746. https://doi.org/10.1111/1541-4337.12606

48. Nakano Y, Tajima M, Sugiyama E, Sato VH, Sato H (2019) Development of a novel nanoemulsion formulation to improve intestinal absorption of cannabidiol. Med Cannabis Cannabinoids 2:35–42. https://doi.org/10.1159/000497361

49. Banerjee A, Binder J, Salama R, Trant JF (2021) Synthesis, characterization and stress-testing of a robust Quillaja saponin stabilized oil-in-water phytocannabinoid nanoemulsion. J Cannabis Res 3:43. https://doi.org/10.1186/s42238-021-00094-w

50. Leibtag S, Peshkovsky A (2020) Cannabis extract nanoemulsions produced by high-intensity ultrasound: Formulation development and scale-up. J Drug Deliv Sci Technol 60:101953. https://doi.org/10.1016/j.jddst.2020.101953

51. Assadpour E, Rezaei A, Das SS, Krishna Rao BV, Singh SK, Kharazmi MS, Jha NK, Jha SK, Prieto MA, Jafari SM (2023) Cannabidiol-loaded nanocarriers and their therapeutic applications. Pharmaceuticals 16:487. https://doi.org/10.3390/ph16040487

52. Lazzarotto Rebelatto ER, Rauber GS, Caon T (2023) An update of nano-based drug delivery systems for cannabinoids: biopharmaceutical aspects & therapeutic applications. Int J Pharm 635:122727. https://doi.org/10.1016/J.IJPHARM.2023.122727

53. Akbarzadeh A, Rezaei-Sadabady R, Davaran S, Joo SW, Zarghami N, Hanifehpour Y, Samiei M, Kouhi M, Nejati-Koshki K (2013) Liposome: classification, preparation, and applications. Nanoscale Res Lett 8:102. https://doi.org/10.1186/1556-276X-8-102

54. Gonçalves RFS, Martins JT, Duarte CMM, Vicente AA, Pinheiro AC (2018) Advances in nutraceutical delivery systems: from formulation design for bioavailability enhancement to efficacy and safety evaluation. Trends Food Sci Technol 78:270–291. https://doi.org/10.1016/j.tifs.2018.06.011

55. Verrico CD, Wesson S, Konduri V, Hofferek CJ, Vazquez-Perez J, Blair E, Dunner K, Salimpour P, Decker WK, Halpert MM (2020) A randomized, double-blind, placebo-controlled study of daily cannabidiol for the treatment of canine osteoarthritis pain. Pain 161:2191–2202. https://doi.org/10.1097/j.pain.0000000000001896

56. Matarazzo AP, Elisei LMS, Carvalho FC, Bonfílio R, Ruela ALM, Galdino G, Pereira GR (2021) Mucoadhesive nanostructured lipid carriers as a cannabidiol nasal delivery system for the treatment of neuropathic pain. Eur J Pharm Sci 159:105698. https://doi.org/10.1016/j.ejps.2020.105698

57. Hommoss G, Pyo SM, Müller RH (2017) Mucoadhesive tetrahydrocannabinol-loaded NLC—formulation optimization and long-term physicochemical stability. Eur J Pharm Biopharm 117:408–417. https://doi.org/10.1016/j.ejpb.2017.04.009

58. Esposito E, Drechsler M, Cortesi R, Nastruzzi C (2016) Encapsulation of cannabinoid drugs in nanostructured lipid carriers. Eur J Pharm Biopharm 102:87–91. https://doi.org/10.1016/j.ejpb.2016.03.005

59. Zielińska A, da Ana R, Fonseca J, Szalata M, Wielgus K, Fathi F, Oliveira MBPP, Staszewski R, Karczewski J, Souto EB (2023) Phytocannabinoids: chromatographic screening of cannabinoids and loading into lipid nanoparticles. Molecules 28. https://doi.org/10.3390/molecules28062875

60. Uttreja P, Karnik I, Adel Ali Youssef A, Narala N, Elkanayati RM, Baisa S, Alshammari ND, Banda S, Vemula SK, Repka MA (2025) Self-emulsifying drug delivery systems (SEDDS): transition from liquid to solid—a comprehensive review of formulation, characterization, applications, and future trends. Pharmaceutics 17:63. https://doi.org/10.3390/pharmaceutics17010063

61. Buya AB, Beloqui A, Memvanga PB, Préat V (2020) Self-nano-emulsifying drug-delivery systems: from the development to the current applications and challenges in oral drug delivery. Pharmaceutics 12:1194. https://doi.org/10.3390/pharmaceutics12121194

62. Laffleur F, Millotti G, Lagast J (2025) An overview of oral bioavailability enhancement through self-emulsifying drug delivery systems. Expert Opin Drug Deliv 22:659–671. https://doi.org/10.1080/17425247.2025.2479759

63. De Prá MAA, Vardanega R, Loss CG (2021) Lipid-based formulations to increase cannabidiol bioavailability: in vitro digestion tests, pre-clinical assessment and clinical trial. Int J Pharm 609. https://doi.org/10.1016/j.ijpharm.2021.121159

64. Kok LY, Bannigan P, Sanaee F, Evans JC, Dunne M, Regenold M, Ahmed L, Dubins D, Allen C (2022) Development and pharmacokinetic evaluation of a self-nanoemulsifying drug delivery system for the oral delivery of cannabidiol. Eur J Pharm Sci 168:106058. https://doi.org/10.1016/j.ejps.2021.106058

65. Knaub K, Sartorius T, Dharsono T, Wacker R, Wilhelm M, Schön C (2019) A novel self-emulsifying drug delivery system (SEDDS) based on VESIsorb® formulation technology improving the oral bioavailability of cannabidiol in healthy subjects. Molecules 24:2967. https://doi.org/10.3390/molecules24162967

66. Sandmeier M, Wong ET, Nikolajsen GN, Purwanti A, Lindner S, Bernkop-Schnürch A, Xia W, Hoeng J, Kjær K, Bruun HZ, Jensen SS (2025) Oral formulations for cannabidiol: Improved absolute oral bioavailability of biodegradable cannabidiol self-emulsifying drug delivery systems. Colloids Surf B Biointerfaces 255:114879. https://doi.org/10.1016/J.COLSURFB.2025.114879

67. Sandmeier M, Hoeng J, Skov Jensen S, Nykjær Nikolajsen G, Ziegler Bruun H, To D, Ricci F, Schifferle M, Bernkop-Schnürch A (2025) Oral formulations for highly lipophilic drugs: Impact of surface decoration on the efficacy of self-emulsifying drug delivery systems. J Colloid Interface Sci 677:1108–1119. https://doi.org/10.1016/J.JCIS.2024.07.233

68. Hermush V, Mizrahi N, Brodezky T, Ezra R (2025) Enhancing cannabinoid bioavailability: a crossover study comparing a novel self-nanoemulsifying drug delivery system and a commercial oil-based formulation. J Cannabis Res 7:35. https://doi.org/10.1186/s42238-025-00294-8

69. Raman Kallakunta V, Dudhipala N, Nyavanandi D, Sarabu S, Yadav Janga K, Ajjarapu S, Bandari S, Repka MA (2023) Formulation and processing of solid self-emulsifying drug delivery systems (HME S-SEDDS): a single-step manufacturing process via hot-melt extrusion technology through response surface methodology. Int J Pharm 641:123055. https://doi.org/10.1016/J.IJPHARM.2023.123055

70. Han B, Shi Y (2025) A content analysis of cannabis edible product characteristics, packaging features, and online promotions. Prev Med (Baltim) 198:108336. https://doi.org/10.1016/J.YPMED.2025.108336

71. EUPVP (n.d.) Common catalogue: varieties agricultural plant and vegetable species. https://Ec.Europa.Eu/Food/Plant-Variety-Portal/

72. EU2015/2283 (2015) Regulation (EU) 2015/2283 of the European Parliament and of the council of 25 November 2015 on novel foods

73. Engeli BE, Lachenmeier DW, Diel P, Guth S, Villar Fernandez MA, Roth A, Lampen A, Cartus AT, Wätjen W, Hengstler JG, Mally A (2025) Cannabidiol in foods and food supplements: evaluation of health risks and health claims. Nutrients 17:489. https://doi.org/10.3390/nu17030489

74. Lachenmeier DW, Habel S, Fischer B, Herbi F, Zerbe Y, Bock V, Rajcic de Rezende T, Walch SG, Sproll C (2020) Are side effects of cannabidiol (CBD) products caused by tetrahydrocannabinol (THC) contamination? F1000Res 8:1394. https://doi.org/10.12688/f1000research.19931.3

75. Lindsay CM, Abel WD, Jones-Edwards EE, Brown PD, Bernard KK, Taylor TT (2021) Form and content of Jamaican cannabis edibles. J Cannabis Res 3:29. https://doi.org/10.1186/s42238-021-00079-9

76. Bonn-Miller MO, Loflin MJE, Thomas BF, Marcu JP, Hyke T, Vandrey R (2017) Labeling accuracy of cannabidiol extracts sold online. JAMA 318:1708. https://doi.org/10.1001/jama.2017.11909

77. FSA (2025) Food standards agency updates guidance allowing CBD businesses to reformulate products on the Public List for safety reasons
78. Scientific Opinion on the risks for human health related to the presence of tetrahydrocannabinol (THC) in milk and other food of animal origin (2015) EFSA J 13. https://doi.org/10.2903/j.efsa.2015.4141
79. Siddiqui SA, Fidan H, Stankov S, Mehdizadeh M, Ambartsumov TG, Kharazmi M, Singh S, Jafari SM (2023) Are cannabidiol (CBD) levels in consumer food products well tested?—a review. Food Front 4:1778–1793. https://doi.org/10.1002/fft2.308